Date: 6/22/15

338.2724 MAR
Martin, Richard,
Coal wars : the future of
energy and the fate of the

D0966256

DISCARDED:
PALM BEACH COUNTY
LIBRARY SYSTEM
OUTDATED 3650 SUMMIT BLVD.
WEST PALM BEACH, FL 33406

COAL
WARS

THE FUTURE OF ENERGY AND
THE FATE OF THE PLANET

RICHARD MARTIN

palgrave
macmillan

COAL WARS
Copyright © Richard Martin, 2015.
All rights reserved.

First published in 2015 by PALGRAVE MACMILLAN® TRADE in the United
States—a division of St. Martin's Press LLC, 175 Fifth Avenue, New York, NY
10010.

Palgrave® and Macmillan® are registered trademarks in the United States, the
United Kingdom, Europe and other countries.

ISBN: 978-1-137-27934-7

Portions of this book appeared, in different versions, in *Fortune, The Atlantic,* and
Sierra magazines. Used by kind permission of the publishers.

Library of Congress Cataloging-in-Publication Data

Martin, Richard, 1958–
    Coal wars : the future of energy and the fate of the planet / Richard Martin.
      pages   cm
    ISBN 978-1-137-27934-7 (hardback)
    1. Coal—Environmental aspects. 2. Power resources. 3. Renewable energy
sources. 4. Coal trade. I. Title.
TD196.C63M37   2015
338.2'724—dc23
                                                                        2014041806

Design by Letra Libre, Inc.

First edition: April 2015

10 9 8 7 6 5 4 3 2 1

Printed in the United States of America.

For the Martins:
Garth, Joann, Doug, and Greg,
and all our children

I had come . . . to see what history meant in flesh and blood. I learned now that it might follow, because an empire passed, that a world full of strong men and women and rich food and heady wine might nevertheless seem like a shadow show: that a man of every excellence might sit by a fire warming his hands in the vain hope of casting out a chill that lived not in the flesh.

—Rebecca West, *Black Lamb & Grey Falcon*

Coal always curses the land in which it lies.

—Henry Caudill, *Night Comes to the Cumberlands*

CONTENTS

PART IV

GROUND ZERO

PROLOGUE

CAPE FEAR

On August 22, 2014, the twin smokestacks at the Cape Fear Steam Plant, a coal-fired power plant in North Carolina, were toppled by dynamite. Meticulously planned for months, the operation was as neat as a chef cutting the legs off a broiler. Matched charges, timed to blow simultaneously, were packed around the feet of the 200-foot-tall brick stacks. When they went off, smoke blasted from the bases and the tops of the towers, nearby sheds disintegrated, and a shudder seemed to travel up them before they fell slowly sideways, in perfect, stately parallel, like a pair of dignified drunks collapsing arm-in-arm in a gutter. The stacks broke apart on impact. Old bricks, blackened by years of coal smoke, flew everywhere. A cloud of dirt and ash covered the scene and settled. Very quickly the noise of the surrounding woods returned and a haze of dust drifted east, toward the coast.

The Cape Fear plant was built in the early 1920s to feed the growing hunger for electricity in Raleigh and Charlotte. It sat between the two cities on the Deep River, an upper fork of the Cape Fear River, which flows south to empty into the Atlantic at Cape Fear, 140 miles downstream. Completed in 1923, it was not a big coal plant: just 320 megawatts or so. During World War II, electrons from the plant flowed onto the grid that fed Fort Bragg, 30 miles away, and the coastal airfields from which spotter planes took off to hunt German U-boats in the western Atlantic. In the 1960s Cape Fear's boilers ran 24/7 to help fuel the postwar economic boom. In the twenty-first century the plant helped light the technology labs and offices of the Research Triangle, North Carolina's high-tech hub. It was reliable, inexpensive to run, and politically uncontroversial. Most people passing on U.S. 1, which crosses southeast North Carolina, never noticed it. And every year it emitted nearly 2 million tons of carbon dioxide into the atmosphere, totaling around 180 million tons over its nine decades of operation.[1]

The plant was owned by Progress Energy, which is a subsidiary of Duke Energy, one of America's largest investor-owned utilities.[2] Progress was in the process of shutting down nine coal plants across North Carolina, eliminating one-third of its coal-fired generation capacity in the state. The company was not shutting down these facilities out of concern for the environment; it was closing coal plants out of a simple economic calculation.

In the last five years America's fleet of aging coal plants has been under attack from two sides: government regulators, who have progressively tightened the state and federal limits on air pollution (primarily mercury, sulfur dioxide, and nitrogen oxides) from power plants, requiring coal plant operators to install hundreds of millions of dollars' worth of scrubbers and other pollution-reducing technology; and the market for electricity, primarily because of abundant supplies of low-cost natural gas made available by America's shale gas revolution. Simply put, coal plants—especially ones built before 1980 that lack most modern pollution-reduction systems—are increasingly uneconomical to run. Many of these plants have far outlived their projected life spans and would require costly upgrades in order to win approval from the Environmental Protection Agency to stay open.

In the case of Cape Fear, these pressures were exacerbated by disastrous levels of water pollution. In February 2014 a pipe ruptured at a Duke Energy storage pond near Eden, North Carolina, and spilled 30,000 to 40,000 tons of coal ash—the toxic residue of burned coal—into the Dan River, "coat[ing] 70 miles of the waterway in North Carolina and Virginia with a toxic sludge that turned the river a sickening gray color," according to an account in RealClearPolitics.[3] It was one of the largest coal ash spills in U.S. history. Barely a month later, Duke was cited for dumping 61 million gallons of wastewater into the Cape Fear River from another ash storage facility. Demolishing and cleaning up the plants themselves will take two or three years. Cleaning up the coal ash accumulated over the plant's long life will take decades, if it ever happens.

Progress, its parent company Duke, and every other major coal-burning utility in the United States is faced with a complex series of decisions over the next ten years: shut the plants down or keep them running for another three, five, or maybe even ten years.

Those decisions will be made in boardrooms, regulatory hearings, legislatures, and courtrooms across the country, as hundreds of coal plants reach the end of their lives. They will determine the future performance of many big companies and the results their shareholders receive. But they are only one part of a much larger struggle over the future of the global coal industry.

IN THE LAST TEN YEARS, without really realizing it, we have embarked on the largest peacetime deconstruction project in the history of the world. It started in the early years of this century with the retirement, on schedule, of a few pre–World War II coal plants. It has accelerated—*metastasized,* if you're a coal industry executive—into a comprehensive program, driven in many countries by market forces and government edict, to dramatically reduce the number of power plants burning coal to make electricity.

The magnitude of the task is overwhelming. In the United States alone, as of 2014, there are about 580 coal plants, most of them concentrated in the East and the Midwest, burning 2.6 tons of coal a day.[4] More than 90 percent of these plants are older than 20 years; some, like the 15-megawatt C. C. Perry K. Steam Plant in Indianapolis, which was built in 1925 and was still in operation as of the fall of 2014, are much older. In 2013 I wrote a report for Navigant Research on the business of decommissioning coal plants, which pulled together several estimates of coal plant retirements over the next two decades, including one from the U.S. Energy Information Administration. Under the EIA's base forecast, 49 gigawatts (or GW; one gigawatt equals 1,000 megawatts) of capacity, out of a total coal-plant output of 340 GW, will be eliminated by 2020. If the price of natural gas remains low, that number will climb to 55 GW of retired coal capacity; in a scenario of low economic growth, retirements could reach nearly 70 GW, or 21 percent of existing capacity.[5]

Many analysts consider those numbers low. A report by the Union of Concerned Scientists found that 353 coal-fired units (one plant often has multiple units), representing 59 GW of capacity, are "ripe for retirement" and may be shut down by 2020.[6] The Brattle Group, an energy consultancy, said in an October 2012 paper that coal plant retirements through 2016 will reduce coal capacity between 59 GW and 77 GW.[7]

Seventy-seven GW by 2016: that's more than one-fifth of the coal burners in America, about the total capacity of power generation in Ireland, to be shut down by 2016. My own conclusion, having examined the data from the last few years and conducted hundreds of coal industry interviews, is that the higher forecasts are likely to be more accurate and that, in any event, the coal shutdown is accelerating.

It will accelerate further if the EPA's proposed rule for limiting carbon dioxide emissions from power plants, known as the Clean Power Plan, becomes law. Announced in August 2014, these rules would allow states to craft individual plans to reduce their carbon output, reducing nationwide carbon emissions by 30 percent from 2005 levels by 2030. The EPA estimates that the plan would eliminate about 730 million metric tons of carbon released into the atmosphere annually, providing between $55 billion and $93 billion in public health and economic benefits on an investment of $7.3 billion to $8.8 billion.[8] The majority of those cuts would come from coal plants. Industry executives and their political supporters have howled that these new rules are the most aggressive thrust in the Obama administration's "War on Coal."

Other countries have already instituted similar regulations, most notably the European Union's "20-20-20" rule: by 2020, the EU member nations are pledged to cut greenhouse gas emissions by 20 percent while upping their use of renewable sources to 20 percent of overall energy consumption and increasing energy efficiency by 20 percent. Even China, the world's largest producer and consumer of coal, has vowed to begin reducing its carbon emissions from coal-fired plants.

It remains to be seen whether these regulations will ever take effect. Republicans have promised a fierce legal battle against the EPA's Clean Power Plan that could delay implementation until after President Barack Obama leaves office. That could doom the plan, particularly if a Republican succeeds him. Ultimately, one way or the other, the decline of the coal industry is irreversible; the only question is how long it will take, and whether it will happen in time to make an appreciable difference in limiting catastrophic climate change.

ALL FOSSIL FUELS EMIT CARBON DIOXIDE when burned, but coal is a uniquely dense source of the greenhouse gas. Bituminous coal, the most common type of coal used for power generation, releases 205.7 pounds of carbon dioxide per million BTU of energy; natural gas releases 117 pounds per million BTU.[9] Coal historically has accounted for around half of the energy used in the United States; today it has dropped to less than 40 percent.[10] Worldwide, coal generation produces 42 percent of the global electricity supply, and just under 30 percent of the world's total energy demand (it's much more in some countries, including China, where 75 to 80 percent of the electricity comes from coal). Coal burning is the single largest source of carbon emissions in the world today, accounting for more than 44 percent of global carbon dioxide emissions.[11]

Since the late eighteenth century, when it emerged as a source of heating and, later, steam power, coal has brought untold benefits to mankind. Coal fired the Industrial Revolution. It powered the ships that raised empires in the nineteenth century, and pierced the darkness and warmed the night in millions of homes. It is abundant, cheap, and spread liberally across the earth's surface. The mines and the coal trains and the stinking power plants are out of sight of most people in the developed world today (a statement that's certainly not true for the citizens of China, the world's most coal-polluted country). Most people barely give a thought to where their electricity comes from when they flick a light switch or plug in a laptop. But our modern technological society would be inconceivable without cheap energy from coal.

And that society will not survive unless we wean ourselves off coal. To limit catastrophic climate change, we must find a way to power our cities with less polluting energy sources, and we must do it in the next couple of decades. For the last century and a half we have made a devil's bargain with coal, and now the payment has come due. If coal consumption is not drastically reduced in the next 20 years, a climatologist told me, "it's game over."

In other words, it doesn't matter how many roofs we cover with solar panels, how many wind farms we build, or how many electric vehicles we sell; if we don't shut down Big Coal, the fight against global warming is lost. This book is about that struggle.

IN THE SPRING OF 2009 I stood beside a country road in northern Tennessee, looking over the worst industrial spill in U.S. history. I was traveling with a pair of nuclear engineers while reporting for my first book, *SuperFuel,* about the renaissance of thorium-based nuclear power. What we were looking at now, though, was a calamity straight out of the nineteenth century.

A lumpy, black porridge stretched away from the road, coating the green fields. It was as if an enormous black scab had formed over a mile-long gash in the earth's surface. Several months before, in December 2008, a containment wall at the Kingston coal plant had failed, releasing a billion-gallon flood of coal ash—100 times the amount of material released in the *Exxon Valdez* oil spill that decimated Alaska's coastline in 1989. Just about every coal plant has a coal ash pond out back. The sludgy, grayish-black material contains a devil's brew of toxins: arsenic, mercury, barium, chromium, and half a dozen others. The Kingston spill released 140,000 pounds of arsenic into the Emory River and turned about 400 acres of verdant farmland into a toxic waste site. It would cost more than a billion dollars to clean up and would lead to years of litigation and millions of dollars in fines for the Tennessee Valley Authority, the owner of the Kingston plant. And it would become a potent symbol of our deadly addiction to coal.

The book I was working on was about an advanced nuclear technology that could become a solution for the world's energy crisis. But looking over the Kingston disaster, it struck me that we are still dependent on a primitive energy technology: burning carbon-laced rocks. As the former mayor of Boulder, Colorado, would tell me years later, "We're using nineteenth-century technology, developed in a twentieth-century regulatory environment, to supply our power in the twenty-first century." The seed of this book was planted that day in Tennessee.

Six years later, progress in shutting down coal is not encouraging. According to the World Resources Institute, nearly 1,200 new coal-fired plants, with a total capacity of more than 1,400 GW (three times the size of the U.S. coal industry today), are proposed or planned in 59 countries. China and India account for three-quarters of those. The developing world is desperate for new energy sources, and coal is the cheapest and most readily available. By some projections, coal use in China could double by 2035. While coal use is on the decline in many countries, most notably the United States, it's increasing in places like Germany and Japan, which both began to phase out their nuclear plants in the wake of the nuclear accident that followed the Fukushima earthquake and tsunami in 2010. Coal use spiked in Germany in 2013 and now accounts for nearly half of its power generation. Coal's decline may be inevitable, but it is putting on one hell of a death scene.

Consider the shipping industry. The first steamships began to transport coal along the coast of England in the early nineteenth century. At that time the dominant mode of seaborne transport was sailing ships. Just before the start of World War I, a century after the advent of steam, there were still sailing ships carrying cargo, some of them out of British ports. It took more than a hundred years to shut down the world's fleet of commercial sailing vessels. If it takes that long to shut down the coal industry, Mumbai, Manhattan, and Miami will be underwater, and modern industrial society will be in some stage of collapse.

This is not a book of policy, though, nor a polemic on the evils of coal. It's a narrative of the front lines. Crafted as a series of journeys—through Appalachia, across Wyoming's Powder River Basin, deep into China's Shanxi Province, to the Yampa Valley of Colorado, and to southern Ohio—it's about the people who find themselves caught up, in one way or another, in the massive transformation that is roiling the energy industry, disrupting companies and communities, and erasing forever a way of life. Although my point of view is evident throughout—either we shut down coal or it will destroy us—the reporting in the chapters ahead is meant to be unbiased and unblinking.

I try in writing and speaking to avoid war metaphors; they are almost always overstated and they detract from the horrors of real war. In this case, obviously, I've made an exception. I justify this because the struggle over the future of coal is a war that is as existential, imperial, and immensely destructive to life and property as the world wars of the twentieth century.

First, at stake in the coal wars is our survival—perhaps not as a species, but certainly as people inhabiting societies and economies that are based on cheap, dirty energy. The battle lines are clearly drawn: on one side are the people working to shut down the industry, and on the other those fighting to preserve it, plus a vast group of interested spectators that, ultimately, includes everyone alive today along with their children and grandchildren.

Like many wars, this one features an empire. The empire of coal has neither a capital nor a single emperor, but rather many lords ruling many fiefdoms. It is global in reach, expansive in nature, and reactionary in politics. It's backed by enormous treasure and armies of cheap labor, and like the Roman, the Ottoman, and the Chinese empires, its urge for self-preservation at all costs will probably keep it running, decaying and diminished, long past its natural end. Historians will one day comb the archives, seeking to pinpoint its apogee and the moment the fall began.

Finally, the coal wars have already claimed casualties. They include not only a Mongolian sheep herder named Mergen who was killed when he tried to block a coal truck at a huge mine in his country in 2011, but also the thousands who still die every year in mining accidents, from black lung and emphysema, and as a result of air and water pollution from the coal industry—as well as those who will lose their homes and their livelihoods to global warming.

There will be many more. I have tried to document them before they fall.

PART I

THE DEATH SPIRAL

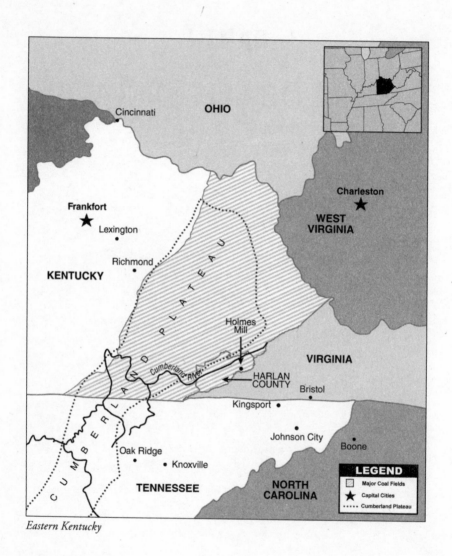

OHIO

Cincinnati

WEST
VIRGINIA

Charleston ★

Frankfort
★
Lexington

Richmond

KENTUCKY

C U M B E R L A N D   P L A T E A U

Cumberland River

Holmes
Mill

VIRGINIA

HARLAN
COUNTY

Bristol

Kingsport

Johnson City

Boone

C U M B E R L A N D

Oak Ridge

Knoxville

TENNESSEE

NORTH
CAROLINA

LEGEND
Major Coal Fields
★ Capital Cities
Cumberland Plateau

Eastern Kentucky

CHAPTER 1

THE TVA

Bill Pritchard had always worked for the Tennessee Valley Authority, until he didn't. Pritchard grew up in Memphis, a few miles from the Allen Plant, a municipal coal plant that was leased and later purchased outright by the authority. He hunted and fished the river country of Tennessee, and in high school he wrestled ("In Memphis you either play basketball or you wrassle," he told me). He stayed home for college at Memphis State, now the University of Memphis. MSU was a basketball school, and it drew its players from the surrounding cities of the mid-South. One of the stars in Pritchard's freshman year was Dexter Reed, a smooth and graceful guard from Little Rock, my hometown, whom I'd grown up watching and a few times trying, unsuccessfully, to guard. Pritchard graduated in 1980, the same year I graduated from college, and he went to work for the TVA the next day. Except for a three-year Wanderjahr when he went off in search of himself and attended graduate school, he's worked there ever since. In those days, when you went to work for the TVA you were usually signing on for life.

An electrical engineer, or double-E, Pritchard joined the authority's training program for engineers, in which he would rotate from plant to plant for three years. His first job was in Hollywood—Hollywood, Alabama, where the Bellefonte nuclear plant was at that time under construction. He was assigned to assist with the final design documentation for the two 1,256-megawatt reactors. This was in the early 1980s, at the peak of the nuclear power construction boom. TVA's nuclear program included Bellefonte plus the two-reactor Watts Bar plant, near Knoxville, as well as the Clinch River breeder reactor, conceived as the nation's first plutonium-based, self-sustaining nuclear plant—the future of power generation.

Things didn't go as planned. "They quickly found they didn't necessarily need all those people at Bellefonte," Pritchard recalls, "though they were still planning on building it at the time."

Reconsideration came quickly. Caught up in the wave of post–Three Mile Island, post-Chernobyl backlash against nuclear power, and plagued with the cost overruns and delays that afflict most nuclear power projects, Bellefonte was never completed. When it was officially abandoned in 1988, Unit 1 was 88 percent complete and Unit 2 half-finished. Six billion dollars and countless man-hours had been invested in the site, and not a kilowatt of electricity was ever generated. By that time Pritchard had moved on; he did a stint at Watts Bar as well, where the second reactor, also incomplete, was shut down in 1988. Pritchard's record of working on completed power plants was discouraging, but when he moved to the coal side of the operation he found his home for the next three decades. He specialized in instrumentation, minding the systems and gauges that kept the plant running and monitored its performance. Something in his voice changes when he talks about coal plants.

"I was lucky enough to get on the big project for Paradise, in central Kentucky," he recalls. "As a just-out-of-school engineer, I was working on the whole control system for them ID fans—they're still there."

"ID" stands for induced draft, and it was an early attempt to use technology to reduce some of the environmental damage, and the effects on nearby communities, of burning coal. Like a bellows on a forge, only in reverse, the giant fans pull air through the boiler, where coal is burned to create the steam that spins the turbines to produce electricity, and vent it to the outside. ID fan boilers replaced conventional pressurized furnaces, which tended to dump ash and coal dust over the nearby countryside. It was satisfying work: Pritchard was helping to clean up America's leading source of power. And the money was good, especially since the young and single Pritchard had little time to spend his salary. But he picked up and left. "I got tired of working 80 hours a week, and I went a little crazy." He quit the job, bought a pickup, and drove it across the country. He saw the West and fell in love with Colorado, but it was not enough to overcome the gravitational force of home. Eventually he made his way back to Memphis and picked up a master's degree, also in double-E, also at MSU. And inevitably, he went back to work for the TVA, doing instrumentation for coal plants. By 2012 he'd been at the John Sevier plant, in eastern Tennessee near Rogersville, for 21 years.

"I told my wife maybe we'd be here two or three years," he says, chuckling. "But it's home now. I've lived here longer than I've lived anywhere, all our kids were born here. We're stayin'."

By the time Pritchard reached his mid-50s his life seemed laid out, as far as he could see. He'd risen to be the plant manager. His kids were raised and going off to college. His pension from the authority was secure. He figured he would work another ten years or so, till Social Security kicked in, and then retire to fish and hunt. TVA had gone through plenty of changes in the last ten years, not all of them reassuring. But he figured it was a stable business to be in. People would always need power. And there was plenty of coal.

Then, one day in 2013, he and his fellow Sevier workers were called into the plant auditorium in the middle of the day for a special announcement.

LIKE MANY POWER PLANTS, the John Sevier Power Station is tucked away, hidden well enough that, unless you're looking for it, you're not going to see it. Burning coal to make electricity is like choosing a presidential candidate: the less the public sees of the actual process, the better. John Sevier sits on the Holston River, in eastern Tennessee, a tributary of the Tennessee River, which rises in the limestone escarpments of the southern Appalachians and curls far to the south, across northern Alabama, before snaking back north to join the Ohio near Paducah, Kentucky. Known as the Mountain Empire, stretching from Roanoke in the northeast to Knoxville in the southwest, these deep forests are intricately threaded by the headwaters of the Tennessee, a labyrinth of hollows and sloughs and sluggish streams walled with dense underbrush: the Clinch, the Powell, the Nolichucky, the French Broad, the Holston. The banks of the Holston, which must be described as "sleepy," bear the traces of past industry like the remnants of a former civilization: ramshackle barns, fallow fields, the ghostly concrete stanchions of a vanished bridge.

In fact, the Sevier coal plant itself is now officially a remnant. Named for John Sevier, a tavern keeper who helped lead the frontier wars against the Cherokee and Chickamauga in the decades following the American Revolution and became Tennessee's first governor, the plant was first fired up in 1957. The coal boilers at Sevier operated continuously for 55 years and were shut down for good in 2012, to be replaced by a sparkling new combined-cycle natural gas plant that sits, literally, next door. It's hard to find a more obvious example of America's energy past confronting its future.

That's especially appropriate here in the Mountain Empire, because this region is served by the Tennessee Valley Authority, the iconic New Deal federal agency that, beginning in the 1930s, brought light and air conditioning and refrigeration to some of America's most benighted communities. For decades the TVA was the economic engine of much of the Southeast and, in the case of villages like Rogersville, the nearest town to Sevier, the only significant place of employment other than fast-food joints and retail chains. Sevier embodies the enlightened transition from coal to low-cost, low-emissions, high-tech power generation fueled by natural gas—but it also epitomizes the wrenching changes the TVA, which operates 11 coal plants across its six-state service territory and is one of the largest consumers of coal in the country, is undergoing. For decades the TVA got the majority of its power from coal. Now, driven by the EPA and the governments of the states in which it operates, the authority is being forced out of the coal power generation business. When I arrived in the fall of 2013, the ripples of that change were spreading in expanding circles across the hill towns to river ports, office parks, county courthouses, statehouses, and corporate headquarters in Atlanta, Nashville, Richmond, and all the way to Washington, D.C.

I was staying in Kingsport, which sits at the confluence of the North and South Forks of the Holston, 30 miles east of Sevier. Along with Johnson City, Tennessee, and Bristol, Virginia, Kingsport makes up the Tri-Cities area on the Tennessee-Virginia border; around half a million people live in the three cities and their surrounding suburbs and villages. From their earliest days, the fortunes of these communities have been tightly bound up with the coal that comes from the rich fields of Central Appalachia to the northeast. Kingsport got its name not from George III but from the boatyard at the confluence of the two forks of the Holston, which served for most of the nineteenth century as the head of navigation on the Tennessee and which was founded by James King, a colonel in George Washington's Revolutionary Army who was present at Cornwallis's surrender. Among Colonel King's accomplishments was the establishment of the region's first ironworks at the mouth of Steele's Creek, one of the numberless creeks that fed the Holston and, in turn, the Tennessee. Built in the 1780s, King's forge was fueled not by coal but by charcoal baked from lumber harvested from the surrounding forests.[1] The first known use of coal to forge iron in Tennessee was not until years later.

Standing guard at the confluence, the Long Island of the Holston was an important staging post for travelers headed across the Cumberland Gap and into Kentucky. In the nineteenth century, by mule and wagon and barge, and later by rail and trucks, coal from Kentucky and West Virginia was moved through Kingsport to be loaded onto bigger barges and floated downstream to the cities to the east.

As small southern cities go, Kingsport has a legacy of progressivism dating back to the early twentieth century. It's one of the "garden cities" designed by the Harvard landscape architect John Nolen, who advocated a balance of open space, greenbelts, and office buildings to limit urban sprawl. One of the largest black high schools in the United States, Douglass High, opened here in 1930 and, despite Jim Crow laws that prevented black teams competing against whites, became an athletic powerhouse in the 1940s and '50s. Douglass was closed in the desegregation movement of the 1960s.

Driving in on Interstate 26, past the usual lineup of fast-food franchises, strip malls, and sprung-up churches, it wasn't easy to detect Nolen's influence; but Kingsport's core, set along the wooded river bluffs, has a certain pre-1960s graciousness about it. This was the boyhood home of the southern historian John Shelton Reed, who famously remarked, "Every time I look at Atlanta I see what a quarter of a million Confederate soldiers died to prevent."

A few hundred of those soldiers died, were wounded, or taken prisoner here, at the Battle of Kingsport, on December 13, 1864, in which 300 Rebels for three days heroically fended off a force of 5,500 marauders under General George Stoneman on their way to lay waste to the farms and towns of western Virginia. Eventually the Rebels surrendered, and the loss of the Kingsport landing essentially cut off Tennessee from barge-loaded supplies, including supplies of coal. Whatever you think those Confederate soldiers died to prevent, the results of the world the Civil War made are easy to see in Kingsport today. The most glaringly visible is the massive Eastman Chemical Company plant at the center of town, whose five massive smokestacks tower over an 800-acre facility where various industrial chemicals—including polymers, acetates, and methanol—are produced. Eastman Chemical has its own coal-fired power plant, a 197-megawatt station with 19 small boilers that emits more than 3.7 million tons of carbon dioxide a year.[2] Eastman is also the site of the country's first commercial coal-gasification plant, opened in 1983 to produce chemicals using synthetic gas, or syngas, from converted coal rather than petroleum.

The coal-to-chemicals facility was designated a national landmark in 1995 by the American Chemical Society, and Eastman has been a major customer for the struggling coal mines of Appalachia. In 2013, though, Eastman said that, like many industrial users across the country, it is getting off coal.

"We've been talking about that decision since about 2008," project manager Jim Amstutz told the *Times-News* of Kingsport. "We thought we'd be doing renovations to the coal facilities but as the price of natural gas has come down, that has made natural gas the preferred option."[3]

Eastman will not forego coal altogether: the syngas plant will continue to operate. The changeover will reduce Eastman's carbon footprint at Kingsport by only about 20 percent. Like many businesses and utilities across the South, though, Eastman has come to the realization that shutting down its coal boilers is more economical, not to mention more politically acceptable, than continuing to run them. The coal shutdown will cost the company around $90 million, but Eastman expects to save money in the long run.

For the TVA, deeply intertwined as it is with the economic and political life of the region, the costs of kicking coal will be much harder to calculate.

THE TENNESSEE VALLEY AUTHORITY is most famous for building dams. But its real story, its core business for most of its 81-year history, has been operating coal plants.

In the era of the Tea Party and *House of Cards*, it's hard to recall the level of idealism that fueled the creation of the TVA in the depths of the Depression. Franklin Roosevelt swept into office in 1933 on a tide of populist fervor and widespread belief that the government could save the economy and right the social injustices that brought on the Crash. Among FDR's first tasks was to overhaul the power sector.

"Never shall the federal government part with its sovereignty or with its control of its power resources while I'm president of the United States," Roosevelt declared in his first campaign for president, tapping into the wave of anticorporate outrage that crested in the 1930s, a time when largely unregulated private utility holding companies, mostly coal-powered, controlled more than 90 percent of the nation's electricity. Created by Congress in 1933, the TVA was to be more than a builder of dams and a supplier of electricity to poor communities across the Southeast; it was to be a vehicle of opportunity, a beacon of social justice, and a model for the development of backward, agricultural, largely lightless societies at home and abroad.

"TVA was, in effect, the first comprehensive and unified effort to harness natural and human resources for productive purposes, within an ideological context of renewal, conservation, and restoration," wrote historian Steven Neuse in an essay commemorating the authority's fiftieth year of operation.[4]

It was also one of the few federal agencies that inspired folk music, paintings, and poetry; it even found its own Virgil in the writer James Agee, who on assignment for *Fortune* penned a 6,000-word prose poem to the new authority shortly after its founding.

The Tennessee Valley was "the laboratory for a great experiment," Agee wrote, and the authority was setting out "to fashion a civilization which, in a certain important way, is new and is significant to all the U.S." TVA's vision, in Agee's telling, was simple yet audacious: "The natural forces and resources in the valley will be developed with one eye on the long future and the other on the immediate welfare of the people."[5]

Opponents countered that the very idea of the TVA—a government agency that would undercut established power providers and essentially take over the economic development of an entire region—was antithetical to American capitalism. The most vocal early critic was Wendell Willkie, the president of Commonwealth & Southern Corporation, one of the country's largest private utilities, which supplied power to much of the TVA territory. Wilkie, who was to become the 1940 Republican nominee for president, declared government-supplied power equivalent to socialism.

In some ways Willkie and his kind were right. The TVA was the epitome of a centrally planned economy, born fully formed from the forehead of technocrats like its first chairman David Lilienthal (later head of the Atomic Energy Agency), who believed that they, far better than local elected officials, could raise up a region that had progressed little since the end of the Civil War. Finally, the TVA represented the dawning of a nascent environmentalism that recognized, however dimly, that long-term prosperity could not be achieved without stewardship of the health of the land.

"Far and wide the opinion—sound, bad, and indifferent—grows that we are approaching a turning point in civilization, that among other things an ancient human habit must be corrected," Agee declared. "Man must learn to cooperate with his surroundings instead of disemboweling and trampling and hoping to discard them. On the crest of this wave of talk and overrapid action, TVA is the first American attempt to tackle the problem specifically and bit by bit to build at the pace which scientific advancement requires."[6]

Needless to say, the Republican business class that supported Willkie had little patience for such rhetoric. Later, Ronald Reagan would be fired from his lucrative spot as the host of General Electric Theater after calling the TVA a pernicious example of big government. And the fact is the authority drove many of the local utilities out of business, often buying up their assets when they couldn't compete; still, in its first two decades, it was hard to argue with the TVA's success.

By the end of World War II, the authority was the largest supplier of electricity in the United States. It had helped fuel the Manhattan Project, supplying most of the electricity for the uranium enrichment facility at Oak Ridge, Tennessee. It had built a series of locks and dams that made the entire length of the Tennessee, from Knoxville to Paducah, navigable; completed one of the largest hydropower construction programs in history; and brought power, electric light, libraries, and modern agronomy to countless small towns across the region. Whole dark spaces on the map of the Southeast now glowed at night thanks to the TVA.

But over time its success outstripped its ability to perform its core mission: supply enough low-cost power to meet demand from its customers, the smaller distribution companies that sold electricity directly to homes and businesses.

The dams were not enough; what the Tennessee Valley needed was more energy. It needed coal.

Starting with the huge Johnsonville plant in 1949, the TVA went on a massive surge of coal-plant construction: Widows Creek (construction started in 1950); Kingston, Colbert, and Shawnee (1951); John Sevier (1952); Gallatin (1953). Smokestacks rather than graceful concrete dams became the authority's signature structures. By 1960 TVA generated more than 72 percent of its power from coal, and still it wasn't enough: the authority simultaneously embarked on the construction of several nuclear plants, most of which would be canceled in the 1980s in the wake of the Three Mile Island accident. Adding huge amounts of coal-fired and nuclear power generation was a complicated step for a federal agency, and the idea of new federal funding to build coal plants was a nonstarter, even in the 1950s, so TVA's chairman at the time, Herbert D. Vogel, sought to give the authority the ability to raise its own money by issuing bonds. In 1959 Congress passed legislation ending federal appropriations for the authority. After 25 years as a federal agency, the TVA was to be self-sustaining.

That has proven an elusive goal. By the mid-1980s TVA had ballooned into a bloated bureaucracy, saddled with major cost overruns at its nuclear plants as well as dozens of unrelated and unproductive business units (including, at one time, a herd of buffalo). Its original charter, to supply the lowest-priced energy possible to as many people as possible, gradually faded as high electricity rates slowed the region's economy. Bankruptcy became a serious possibility. The solution of President Reagan, never a fan of the authority, was to hire a hatchet man.

Appointed chairman in 1988, Marvin Runyon (a longtime auto executive who had risen to prominence running Nissan's manufacturing plant in Tennessee) immediately pledged that electricity rates would not rise during his tenure. To fulfill that promise, Runyon promptly set about jettisoning nonessential businesses and slashing costs, largely through layoffs: on a single day in June 1988, Runyon jettisoned 5,700 employees and another 1,800 contractors. Over the next few years TVA's workforce was cut in half. Runyon's slash-and-burn management style helped ensure the authority's survival, but it also traumatized the remaining employees and forever altered the authority's

paternalistic relationship with the people of the region. "Renewal, conservation, and restoration" gave way to a new rigor. The progressive vision of its
Depression-era ideals a distant memory, TVA was now just another utility.

AND THAT IS A TOUGH BUSINESS to be in in the twenty-first century. Once a
clubby, comfortable industry in which electricity rates were set by public utility
commissioners who tended to be golfing buddies of the utility executives they
oversaw (and who in many cases were themselves former utility executives), the
business of supplying power—especially from outdated coal plants—is in the
process of transformation. The growing use of renewable energy sources, especially from distributed sites including rooftop solar arrays, has put traditional
utility business models at risk. A general understanding of the power utility
industry in this country is necessary to grasp the upheavals these companies
are facing as the coal era wanes.

In most developed countries, until recently, the generation and sale of
electricity has been a state-owned monopoly. In the United States there have
always been two kinds of power utilities: private, or investor-owned utilities,
known as IOUs, and public utilities, owned by governments or municipalities.
Very few public utilities own their own generation facilities, i.e., power plants;
they are usually distributors, not producers, of electricity. In the tension between these two models, and the incentives that drive them, lies much of the
dysfunction of today's power industry.

The career of Samuel Insull, the British-born electricity magnate who cofounded the companies that would become General Electric and Commonwealth Edison, embodied those tensions. A protégé of Thomas Edison, Insull
came to the United States in 1881, at 21, and finagled his way into becoming
the great inventor's personal assistant. Over the next decade he consolidated his
power as Edison's chief executive, becoming president of Chicago Edison in
1892. Insull proved to be an innovator and an early adopter of new technologies: he pioneered the use of variable electric rates based on time of use and
the load on the system, and Chicago Ed was the first utility to gamble on the
new steam turbines that replaced older reciprocating engines as the mechanism
for turning heat and steam into power. In 1907 he merged two of his utilities
to form Commonwealth Edison, the nation's largest supplier of electricity. By

1920 Commonwealth Edison employed 6,000 people, burned more than 2 million tons of coal a year, and supplied 500,000 customers across the Midwest, earning $40 million in annual revenue.

Insull was also an innovator in the complicated corporate structures that shielded investors from risk and prevented government regulators from grasping, much less effectively overseeing, the business of supplying power. The result was a highly centralized, highly leveraged industry whose inner workings were opaque to outsiders. By 1932, Comm Ed and seven other large utility holding companies controlled 73 percent of the investor-owned electric industry.[7] Because these companies tended to be located outside most of the states in which they operated, they were beyond the reach of state regulators. Early on in the Depression, Insull's holding company—which at its peak controlled $4 billion in assets although his actual equity was only around $27 million—collapsed, wiping out the life savings of 600,000 Americans.

Around the same time, coal mining, never a gentleman's game, reached the apex of its lawless interwar heyday.

"By March 1933 the coal industry, by over-expansion, price-cutting, broken-down wage rates and living standards, and a mounting number of unemployed miners, had contributed more than its share to creating the crisis which confronted the incoming Roosevelt administration," wrote Merle Vincent, a Denver lawyer and muckraking author.[8]

FDR's solution was the Public Utility Holding Company Act of 1935, a signal piece of New Deal legislation. The act forced utility holding companies to spin off unrelated, unregulated businesses; simplified their structure to eliminate multiple layers of ownership; and limited IOUs to smaller geographic areas. Insull himself fled to Europe to escape prosecution, but was later extradited from Turkey and stood trial on mail fraud and antitrust violations. Although he was acquitted on all counts, his lordly career as an influential tycoon was over. Granted a pension of $21,000 a year from his former companies, he led a nomadic existence until he collapsed in a Paris metro station in 1938. On his death he had 30 francs and a laundry ticket in his pockets.[9]

The public utility bill curbed the worst abuses of the holding companies, and over the next two decades the TVA and its sister organization, the Rural Electrification Administration, brought electricity to thousands of poor households. But these measures hardly solved the problems of concentrated ownership and effective monopolies in the power sector. The energy crisis of the 1970s caused retail electricity rates to soar and drove utilities to embark on a huge program to build new power plants that burned domestic fuels—mostly coal and uranium. As the capital costs of all that construction rose, regulators approved further hikes in electricity rates, leading to a vicious cycle of price increases and overbuilding, and sending several big IOUs to the brink of bankruptcy. "Rate cases"—the process by which utilities applied to regulators for price hikes to businesses and consumers—became drawn-out and contentious. Utility lawyers profited, but electricity users (dismissively referred to as "ratepayers" or, in the immortal phrase of Witt Stephens, the Arkansas tycoon who made his first fortune at Arkansas Louisiana Gas Company, "biscuit eaters") did not. By the 1990s it was clear the system was broken.

Utilities, as large corporations have throughout the history of capitalism, responded with a wave of consolidation and acquisition that created an industry landscape similar to that of the reviled holding companies of the 1920s, where a few big power companies control power generation and wholesale electricity transmission across multistate regions. Two examples will provide a sense of the Byzantine complexity of these corporate structures:

Formed from the 1998 merger of Union Electric and Central Illinois Public Service Company, Ameren Corporation is a holding company that in 2003 acquired CILCORP, itself a holding company for Central Illinois Light Company. The following year Ameren acquired Illinois Power, and in 2010 merged its three Illinois subsidiaries into one entity, Ameren Illinois. Based in St. Louis, Ameren now serves 3.3 million customers across a territory of 64,000 square miles. Its CEO, Thomas Voss, made $6 million in 2013.

Holding company American Electric Power, better known as AEP, evolved from American Gas and Electric Company, which was formed in 1906 from the original Electric Company of America. Today AEP comprises a bewildering array of subsidiaries that includes Appalachian Power, AES Ohio (formed from the 2011 merger of Columbus Southern Power and Ohio Power), Indiana Michigan Power, Kentucky Power, Southwestern Electric Power Company, and Wheeling Power, plus AEP Texas, which includes Central Power & Light (doing business as AEP Texas Central) and West Texas Utilities (AEP Texas North). AEP, which sells power to 5 million customers in 11 states, owns the nation's largest electricity transmission system, nearly 39,000 miles of power lines. CEO Nick Akins made $10.6 million in 2013.

Samuel Insull would feel completely at home in these boardrooms. The reforms that FDR fought for in the 1930s have gone the way of the steam engine and the coal scuttle.

These corporate behemoths are built to throw off cash, minimize risk, and stifle competition. But they are ill-constructed to innovate and prosper in the post-coal era. Twenty years from now, 2013 will likely be seen as the year when the traditional utility business model suddenly became obsolete. In an August 2013 story in *BusinessWeek*—headlined "Why the U.S. Power Grid's Days Are Numbered" and based, in part, on data from Navigant Research, the company I work for—David Crane, the CEO of NRG Energy—a large power generator that serves 11 states with both retail and wholesale power generation and distribution—declared that new technology, low-cost natural gas, and distributed renewable energy resources pose "a mortal threat to the existing utility system": "Natural gas is already wiping out coal, and it's going to wipe out most nuclear."[10]

Coming from the head of a company that itself still relies mainly on coal-fired power generation, Crane's comments sent a wave of fear and trembling across the industry. But the fact is that they were hardly news. In January 2013 the Edison Electric Institute (EEI), the prestigious Washington, D.C.–based trade association for the power sector, warned that current trends place modern utilities in the same position as the airlines and the telecom service providers in the 1970s: facing "disruptive challenges" that could place their business models and their very survival at risk. "U.S. [telecom] carriers that were in existence prior to deregulation in 1978 faced bankruptcy," the report observed. "The telecommunication businesses of 1978, meanwhile, are not recognizable today."[11]

The changes rippling across the power sector "all create adverse impacts on revenues, as well as on investor returns," the EEI report continued. "Left unaddressed, these financial pressures could have a major impact on realized equity returns, required investor returns, and credit quality. As a result, the future cost and availability of capital for the electric utility industry would be adversely impacted."

That is analyst-speak for "go broke."

Today's major utilities have three main characteristics that make it difficult for them to adapt to rapidly changing conditions in their business: One, they have huge sunk costs in not only power generation (particularly aging coal plants), but also a rapidly decaying power grid; two, they are in the business of selling electrons, not providing tangible, desirable services (in other words, the more power consumed by people and businesses, the better the utilities fare; when those customers consume less, and start producing their own power, the utilities' business model rapidly erodes); and three, they are reliant on economies of scale. The cost to operate all those coal plants and all those transmission lines rises, inexorably, over time, and as more and more utility customers generate their own power, those costs will be spread across fewer and fewer customers. Every rooftop solar array is another few dollars a month out of the pockets of the power companies. That means that electricity rates must rise to cover the utilities' costs, thereby making distributed, off-grid generation more attractive, causing more people to invest in their own miniplants. This is the "death spiral" that David Crane and others foretell: "Utilities will continue to serve the elderly or the less fortunate, but the rest of the population moves on."[12]

Staving off the threat of distributed renewables has become the primary mission of many utility executives. That rearguard action has mostly taken the form of trying to convince regulators to slap fees on solar-equipped customers who are taking advantage of "net metering"—essentially, the ability to run electricity meters backward so that customers with private generation capacity can get credit for power they supply back onto the grid. In sun-drenched states like Arizona and California, which have the highest rates of small-scale solar adoption in the country, utilities are spending millions in a campaign to persuade politicians to slap what's essentially a solar surcharge on individual homes and businesses with net metering.

"As more customers install solar on their homes, it becomes even more important that everyone who uses the grid shares in the cost of keeping it operating reliably for the future," Don Brandt, chairman and CEO of Arizona Public Service, the state's largest utility, told *SmartGrid News*.[13]

As the price of renewable energy falls (the cost per kilowatt-hour of solar power from commercial scale plants has fallen by more than half since 2010), though, the reality of the situation has become harder and harder to evade:

utilities must reduce their reliance on big centralized coal plants and find ways to adapt to a rapidly changing energy environment where generation is distributed, customers have a choice of providers, and new competitors, from industries far less hidebound, emerge seemingly daily.

"The technology and energy sectors will no longer simply be one another's suppliers and customers," predicts a report from the investment firm Sanford Bernstein. "They will be competing directly. For the technology sector, the first rule is: Costs always go down. For the energy sector and for all extractive industries, costs almost always go up. Given those trajectories, counterintuitively, the coming tussle between solar and conventional energy is not going to be a fair fight."[14]

THE TVA, AS IT HAPPENS, has a higher percentage of the elderly and the less fortunate in its service territory than the utilities of the Sunbelt. In the damp, often cloudy states of the mid-South, solar power and distributed generation pose less of a threat than in Phoenix or Los Angeles. In that sense the TVA has been insulated from some of the turmoil roiling the power sector. In other ways, though, the authority is being shoved into the new world at a much faster rate than utilities in other parts of the country. Partly that's because of its unique, quasi-governmental status, and partly because it found itself on the losing end of a multiyear, precedent-setting court battle over the fallout from burning coal.

On December 22, 2008, when a retaining wall burst at the TVA's Kingston Fossil Plant (not to be confused with the town of Kingsport, farther to the east) in central Tennessee, spilling 5.4 million cubic yards of coal ash across the surrounding countryside—the scene I viewed in 2009—it finished the job that Marvin Runyon had started: turning the TVA's image from that of a benevolent New Deal agency bringing light and progress to the hills to that of a rapacious power utility unconcerned with environmental damage and social costs.

As it turned out, the amount of material that covered close to 600 acres of Tennessee flatland was more than the amount the pond held, according to TVA's official estimate, before the accident. It was the largest toxic spill in U.S. history. Months later the authority would concede that the spill released about 1.09 billion gallons of water and coal ash. High levels of lead and thallium contaminated nearby streams; a train delivering coal to the plant became stuck in the oozing ash; 12 homes were inundated and had to be abandoned; and the ash spilled into Watts Bar Lake, a popular picnicking and fishing spot not far from the plant. The disaster cemented the TVA's noxious reputation in the public mind, and it represented a highly visible episode in the long struggle between environmentalists and regulators who were attempting to come to grips with the effects of coal burning across not only the Southeast, but the nation.

The pond at Kingston was "only one of more than 1,300 similar dumps across the United States—most of them unregulated and unmonitored—that contain billions more gallons of coal ash and other byproducts of burning coal," the *New York Times* reported.[15]

The authority's clumsy response made it seem as detached from the real-world consequences of its actions—and as devoted to its own self-preservation—as any oil company. A spokesman named Gil Francis Jr. said the ash "does have some heavy metals within it, but it's not toxic or anything."[16] The TVA's inspector general concluded in a 2009 report that an analysis of the spill by an outside engineering firm was manipulated to reduce the authority's liability, and "it appears TVA management made a conscious decision to present to the public only facts that supported an absence of liability for TVA for the Kingston spill."[17]

As it turned out, the fallout from the Kingston spill was not the TVA's major legal problem. The coal ash that oozed across the countryside was obvious; the smoke that poured from the stacks at TVA coal plants dissipated quickly but reached farther and, ultimately, carried more extensive consequences for the authority's future.

Dogged for years by complaints that it was violating federal clean air statutes, the authority signed a compliance order with the EPA in 1999 that called for a series of cleanup measures. Those didn't go far enough, and in 2006 the state of North Carolina—which includes six counties, covering more than 5,500 square miles, served by the TVA—filed suit in federal court charging the authority with air pollution violations that endangered the public health. Joining the North Carolina attorney general's office in the suit were the states of Alabama, Tennessee, and Kentucky, along with the National Parks Conservation Association and the Sierra Club. New York and 15 other states filed briefs in support. It was a historic attempt to hold a federal agency that had been producing power from coal for more than 50 years accountable for the transborder pollution associated with burning coal. And it represented a turning point in America's history of coal-fired power generation.

Initially, the plaintiffs won a ruling in federal district court. But TVA appealed, and in July 2010 the Fourth Circuit Court of Appeals overturned the ruling. TVA lawyers argued that the authority was simply doing what it had always done, what its charter called for it to do: provide the lowest-cost power possible to as many people as possible. The Clean Air Act did not specifically outlaw the forms of pollution the TVA was charged with. The North Carolina suit, wrote Judge J. Harvie Wilkinson III, was based in "vague public nuisance standards" that raised the "potential for chaos" and for the "disruption of expectations and reliance interests" of utilities acting in good faith. Victory for the plaintiffs would undermine the "carefully created" legal regime established with the Clean Air Act.[18] For the moment, the TVA could go on burning coal.

North Carolina attorney general Roy Cooper, displaying unusual spine for a Democrat in a mostly red state, appealed to the Supreme Court. "At issue is whether people in North Carolina will continue to die as a result of TVA's excess emissions—even though these deaths could be readily averted by installing and operating modern pollution control equipment," Cooper wrote in his petition to the high court. Having already spent dozens of millions in its defense—and facing hundreds of millions in cleanup and legal fees from the Kingston spill—TVA CEO Bill Johnson had had enough. On April 14, 2011, the TVA signed one of the largest pollution control agreements in U.S. history, agreeing to shutter 18 of its 59 coal generation units at three plants: John Sevier and Johnsonville in Tennessee and Widows Creek in northern Alabama. All ten units at the enormous Johnsonville plant, the biggest in the TVA system, would be shut down. Retiring 2.7 gigawatts of coal-fired power generation, the consent decree—which resolved the states' lawsuit as part of a wider agreement with the EPA—also requires the TVA to install modern scrubbers and other pollution controls at its remaining coal plants, and to accelerate the building of solar and wind power plants. Estimated cost to TVA: $3 billion to $5 billion.

The deal represented "a game changer for how we power our homes and businesses in the Southeast," said Mary Anne Hitt, who heads the Sierra Club's Beyond Coal campaign. But it did more than that: it established the legal power of states to seek redress for the indirect, long-range effects of coal power. And it changed forever the economic calculus for utilities burning coal.

THE JOHN SEVIER COAL PLANT lies at the end of a private road that runs between colonnades of towering pines. Twin smokestacks tower over the boiler building and the turbine hall. Built in 1952 and '53, the plant has that characteristic blend of postwar idealism and utilitarianism that appears so dated today; the architecture announces clearly that the plant was built in a simpler and more naive time. Next door, across a gravel service road and the rail spur that for six decades fed the plant with coal, the new natural gas plant looks futuristic, as if a particularly complicated spaceship had unexpectedly landed. The day I visited was overcast and spitting rain. The new gas plant, silver and gray and white, gleamed dully against the fall sky.

In the plant's conference room I met with the men who were overseeing the shutdown of the coal plant and the start-up of the gas plant: Bob Dalrymple, a regional manager for TVA; Bill Pritchard (still employed although the coal boilers had already gone cold); and Terrell Slider, the manager of the new gas plant, along with a couple of TVA PR men from Knoxville. The changeover at Sevier, said Dalrymple, presented the authority, a symbol of twentieth-century progress that had clung too long to its former business model, with an opportunity to remake itself for the twenty-first century.

"The settlement allowed us to outline a path forward," Dalrymple explained. "It gave us some certainty in direction on a few things that had been unsettled for a while." The authority's overall mission had not changed, but TVA had to adapt to a dramatically shifting regulatory and economic environment. "We have to maintain a balanced portfolio for people of the Tennessee Valley region, and as we evaluated the current state of environmental regulations, the potential future implications for those regulations, what we know now and what we can anticipate, and how we can continue to provide low cost power for the valley, it became clear that continuing to operate this plant wasn't the way to go."

By August 2012, Sevier's four coal units stopped burning. At the time it was thought that Units 3 and 4 would be fired up again—when the dust had settled, the lawyers had been paid, and new pollution scrubbers installed. With billions invested in coal plants that had faithfully produced power for decades, authority veterans couldn't comprehend shutting down more coal units than the consent decree would force them to.

Within a year it became clear that continuing to operate the other two units didn't make sense, either. This was at the dawn of the era of cheap natural gas, and suddenly coal plants that looked like viable long-term producers 12 months before no longer made economic sense. It was no contest. Retrofitting a major plant like John Sevier could cost upward of $1 billion; you could build a new natural gas plant that burned fuel now cheaper than coal (on a per-BTU basis), with half the carbon output, for a little more than half of that.

"The potential was there to restart 'em," Dalrymple told me, sounding almost wistful. "We kept the door open to evaluate our flexibility, to make sure we made the best business decision. But it wound up we couldn't wait."

It was an inevitable decision. It would take years for the full transition, but coal's days as the major producer of baseload power for the Southeast were now numbered. Still, the decision took the Sevier workforce by surprise. Herded into the plant's small auditorium, they took in the news with stunned silence. One hundred seventy people worked at the coal plant. Automated to the hilt, and with no coal to transfer and load and clean up after, the new gas plant would only require 35 full-time employees. The economic engine of this slice of eastern Tennessee was about to disappear.

In microcosm, the shift at Sevier represented what was going on across the TVA. Not only is the authority moving to cleaner, more sustainable, less labor-intensive forms of energy, but also economic growth, and thus power demand, across the valley is nearly flat. At the same time, against skeptics' predictions, energy conservation is taking hold. TVA is producing less power, from plants that require fewer people, for customers who are using less energy. That means fewer jobs.

"The total employment of TVA—well, that's a question that's still being worked on," said Dalrymple. "There's multiple drivers. There's some facility idling that will reduce the number of folks we've had from a historical standpoint. But at the same time we're driving for efficiency and effectiveness, how do we operate the business better. There's multiple things moving to reduce the number of workers at our coal and gas facilities."

TVA is in another kind of spiral: because its plants are in many areas the primary economic engine of the surrounding regions, laying off workers reduces economic activity, thus reducing demand for power. Moving off of coal, becoming more energy-efficient, automating plants so they require fewer workers: these are all good things, in the long run, for society. But they carry unavoidable short-term costs for local economies.

The combination of a venerable federal agency in upheaval, rural economies headed south, lost jobs, and a controversial environmental policy presented a scenario ripe for demagoguery, and as usually happens, up stepped a man fitted perfectly for the role.

Kentucky Republican Mitch McConnell, at that time the Senate minority leader, had a long and vocal record of climate change denial ("As recently as 30–35 years ago we were worried about the globe getting too cold"[19]) and of leadership in the angry opposition to President Obama's so-called War on Coal. In October 2013, McConnell summoned TVA president Bill Johnson to an audience in his chambers in the Russell Building, in D.C., and let him know that neither McConnell himself nor the people of Kentucky would stand by while TVA abandoned its coal plants.

"We are well aware of the pressure that public utilities face from the Obama Administration to transition from using coal to alternative sources of energy," McConnell wrote in a letter to Johnson following their meeting. "To allow a historically abundant and proven resource, such as coal, to fall by the wayside would ultimately threaten our energy independence. Accordingly, we request the TVA maintain existing coal-fired operations—that utilize Kentucky-produced coal stocks—as your agency considers alternative sources in conjunction with coal for electricity generation."[20]

In fact, it is economic forces, more than pressure from the Obama administration, driving the TVA off coal. TVA had already begun shifting away from Kentucky coal, bringing in cheaper supplies from the Powder River Basin in Wyoming by train. That didn't stop McConnell from claiming credit, a few weeks later, for TVA's decision to continue burning coal at one of its three units at the Paradise Fossil Plant in Drakesboro, Kentucky, another plant switching over to natural gas. "I fought hard to prevent these changes and fortunately one of the units will continue to burn coal, saving hundreds of jobs," proclaimed the senator. "I'll continue to fight wherever the administration's anti-coal agenda threatens the livelihood of Kentuckians and I'll fight to get the lost jobs back."

That's not true either. McConnell's intervention hadn't prevented anything; the decision to keep burning coal at Paradise's Unit 3 had been made months earlier. "Let's just say Unit 3 was never in danger of closing," a TVA spokesman drily told Louisville radio station WFPL.[21] Recast as Obama's War on Coal, the industry's inevitable decline gave Republicans a stalking horse for resisting newer and cleaner technologies. But the brooms of their rhetoric couldn't hold back the tide of change, and their support for the industry was of little help to the TVA employees who walked out of the Sevier plant's auditorium after the decision on the plant's future was announced and stared into the bright glare of their uncertain futures. There was to be no restart; the coal boilers had gone silent for good.

"The actual announcement was a shock to all of us," says Pritchard, shaking his head. "We had no idea that the negotiations were going on with the EPA. Nobody here did."

Pritchard, in fact, led the team that had been working on the new scrubbers that were to be installed on Units 3 and 4 to bring them into compliance with the new EPA restrictions on mercury and sulfur dioxide. Sevier was to be the TVA's next big plant retrofitted with modern pollution control systems. It was another in his multiyear history of dead-end projects. "We were working on that for two or three years—they told us we were off the books for a while." The new pollution controls would cost $550 to $600 million. A new superefficient natural gas plant would cost the same, or less. Eventually Johnson and his managers looked clearly at the numbers. Sevier's fate, and that of the 170 people working there, was sealed by economic forces that reached far beyond the Tennessee Valley.

Reducing the authority's workforce is only one of the wrenching changes that Bill Johnson faces. The entire generation mix must be recalibrated and rebalanced as the big coal plants that had been the backbone of the system get powered down over the next five years. The layoffs, though, were the most visible consequence, and here it must be said that TVA has tried to live up to its lingering reputation as a benevolent arm of government rather than a ruthless corporation.

"I feel like TVA has done us a favor," Pritchard says emphatically. "They went out of their way to really make it easy on people to continue working for 'em."

Each employee was asked to put in writing their top three choices, including continuing to work for the authority or taking voluntary retirement. Pritchard's top choice was to continue working on instrumentation on other plants while still living in Kingsport. "That didn't fly." He had two options: move to Chattanooga and join the authority's in-house engineering group, or take early retirement. He chose the latter, although in a paradoxical stroke of good luck he has been asked to stay on, on contract for another six months, to help with the transition. The day we met he'd been scheduled to interview for a new job with a company in Chicago. He canceled that meeting when he got the extended offer from TVA.

"I'm lucky—I've got skills that are transferable. It's all instrumentation and control systems. They'd like you to be a lot younger, though—they think us old-timers don't understand the newer technology.

"There's others, though, that aren't able to move on. They don't have that choice."

Several former Sevier workers are now commuting cross-state, to the Kingston and Bull Run plants, a 200-mile drive. Some have gone in together to rent homes in that area and see their families only on the weekends. "TVA is one of the premier jobs in this area of Tennessee," Pritchard says. "Once you get onto that you really don't want to let go."

The larger economic costs to the region of shutting down the TVA's coal fleet have not been calculated, but the loss of hundreds of jobs, out of a total workforce of nearly 13,000, is a minimum effect. In 2012 the TVA said it would cut at least 1,000 jobs over the next few years to try to rebalance its books and to adapt to the new era of power generation.

The authority, said executive vice president Janet Herrin, will get "about $100 million in labor savings as we work to reduce costs to meet the challenges of the lower revenues we are currently experiencing." The job cuts "will help us to maintain financial health in the near term, while improving competitiveness over the longer term."[22]

Beyond the U.S. Congress, outcry has been surprisingly muted, in part because of TVA's proactive approach to finding new roles for coal plant workers and offering generous settlements for those retiring "voluntarily." Still, a town like Rogersville, with its discount chain store and its off-brand fast-food joints, is unlikely to thrive with the number of jobs at Sevier slashed by 80 percent. As for Pritchard, he's too young and not wealthy enough to actually quit working. He plans to get more serious about the job search once his current contract winds up. He and his wife are committed to staying in East Tennessee. But that might not be their choice.

Before we walked over to the new gas facility, he gave me a tour of the coal plant, now silent and eerily clean. It looked like everything had just been swept, mopped, and polished. In the old control room, the cockpit of the plant, one worker watched over the gauges that monitored that plant's water systems, still operating in case of fire. The analog dials and gauges, which once measured every pound of coal and BTU of heat and kilowatt of electricity for the plant, were darkened. They were redolent of their era, the 1950s and '60s. In the vast turbine hall, the turbines, sheathed in steel, lay like beached humpbacks along the football field–sized floor. Where the air once thrummed with the sound of hundreds of megawatts being generated, you could hear people speaking in a normal tone all the way across the room. Just as quiet was the machine shop.

"This is where it really hits me that it's not coming back," Pritchard said, "when I walk past these doors and it's just dead silent."

And when will the old plant be demolished? "I don't know, they ain't said yet. That'll be a big job, taking these buildings down and cleaning it all up."

IT'S A JOB THAT WILL BE TACKLED many times, not only in the Tennessee Valley but across the United States. Tighter environmental regulations, low-cost natural gas, and the growing anti-coal movement will reduce the number of coal-fired power plants dramatically over the next decade.

In 2012, there were 589 coal plants in the United States, representing about 1,300 coal-fired units and 340,000 megawatts. That was about 30 percent of total U.S. power generation capacity, although in terms of output, because so many coal plants are used continuously as baseload generation, coal generally represents a greater share of the U.S. fleet.[23] Ten years from now, there will be many, many fewer plants burning coal. Some will be shut down and, eventually, demolished; some will be repowered with other fuels, mostly natural gas; some, conceivably, will become shopping and retail complexes, even apartment complexes.

Battered by cheap, abundant natural gas on the one hand and tightening pollution controls on the other, owners of aging coal plants, like the TVA, are finding that the best option is often just to shut them down. By the end of 2013 operators had announced the closure of more than 150 coal plants over the next few years, and the coal generation service at Navigant estimates that up to 45 gigawatts of coal capacity, representing more than 300 units, could be shuttered by 2020.[24]

The most extreme forecast for coal shutdowns came from researchers at Duke University's Nicholas School of the Environment, which concluded that up to 65 percent of the U.S. coal fleet could find itself threatened with shutdown in the next several years.[25]

Sixty-five percent. That would be 383 plants. With an average employment per plant of about 150 people, that's 57,450 jobs. That's not counting the railway workers, machinists, truck drivers, engineers, and other workers whose livelihoods depend directly on the production and consumption of coal. No doubt many of these will be absorbed by other plants—remaining coal plants, or natural gas plants, or other types of facilities. Overall employment in the energy industry is expanding. But many, like the laid-off workers at John Sevier, will be forced onto a job market that is unforgiving for middle-aged workers whose only experience is with a power technology that is soon to be as obsolete as a watermill.

"We're going to need to give special care to people and communities that are unsettled by this transition—not just here in the United States but around the world," said President Obama in his landmark climate change address in June 2013. "And those of us in positions of responsibility, we'll need to be less concerned with the judgment of special interests and well-connected donors, and more concerned with the judgment of posterity."[26]

The judgment of posterity has not been kind to America's record of worker transition in other periods of economic upheaval. From the early 1970s to 2000, employment in the U.S. steel industry dropped from around 2.5 million to fewer than 1 million. A drive through Bethlehem, Pennsylvania, or Gary, Indiana, provides a clear view of how ineffectively the national economy has absorbed, and provided for, those cast-off workers. Steel's downfall was brought on by technological advances, cheap competition from Asia, and structural changes in the global economy that made U.S. steel producers uncompetitive. Coal is different. In the case of the coal industry, we are making a conscious decision to downsize a still-vibrant industry that is sheltered from overseas competition on the generation side (it's hard to sell Chinese electricity to American users) and still competitive on the production side (major U.S. producers like Peabody Energy and Arch Coal are among the most efficient coal producers in the world, and their coal reserves are first-class). It stands to reason, then, that the economic effects of the transition should be more conscious as well.

"Unfortunately, 'transition assistance' in the past has often meant little more than a funeral for workers and communities threatened by the side effects of globalization, environmental protection and other public policies," wrote Joe Uhlein, the former secretary-treasurer of the AFL-CIO's Industrial Union Department, who is now the executive director of the Labor Network for Sustainability.[27] Speaking at a "Good Jobs, Green Jobs" conference in Washington, D.C., Carl Wood, the director of regulatory affairs of the Utility Workers Union of America, put it more graphically: "Workers are used to being ground up and spat out by any change in society. In the United States there is no safety net for the victims."[28]

Anti-coal activists are certainly wrestling with this dilemma.

Coal Free Massachusetts, a coalition dedicated to phasing out coal plants in the state by 2020, has called for a realistic and substantive program to find new employment for the workers whose jobs will evaporate when those plants are shuttered. Specifically, the organization is searching for viable solutions for workers affected by the closure of the huge 1.5-megawatt Brayton Point Power Station in Somerset, Massachusetts. "In Somerset, those who have borne the devastating health burdens caused by coal are now threatened, economically and socially, by the disintegration of jobs and revenues that have funded the basic operations of their town," said a 2012 report produced by Coal Free. "Finding a solution for the future—a just transition—means resolving a tripled conundrum of health, wealth and workforce."[29]

So far the resolution to that conundrum has proved elusive but not unattainable. In Arizona, Southern California Edison closed the Mohave Generation plant in 2005, before the start of the current wave of retirements and closures that is transforming the power sector in this country. A coalition called Just Transition was formed to help reach a deal between the utility and the Hopi and Navajo Indian tribes on whose reservation the plant was located and who supplied the majority of the workers at the plant. With the plant's retirement, SCE found itself with sulfur pollution credits, accumulated under the Acid Rain Program of the Clean Air Act, which it no longer needed and could sell to coal plant operators in other regions of the country. Hopi and Navajo advocates, supported by environmental organizations, argued before the California Public Utilities Commission that the plant had not only supplied a primary place of employment for the tribes but had polluted their lands and consumed vast amounts of groundwater to run a coal slurry from the nearby Black Mesa mine. In 2013 the CPUC ruled that revenue from the sale of the pollution credits would help fund renewable energy projects that will benefit the tribes.

Similarly, in Washington State, a long-running struggle over the fate of the state's only coal-fired power plant, the Centralia Power Plant, resulted in a landmark agreement to phase out the plant's boilers over 15 years while providing $55 million (paid for by the plant owner, Alberta-based TransAlta) for community development and energy-technology transition funding. The circumstances in Arizona and Washington were unique, and not all of the plant closures will have such encouraging outcomes. Harnessing energy to produce power is a messy business, and any transformation as prolonged and far-reaching as the end of the coal era will have both winners and losers. Unfortunately, many of the losers will be residents of the region that was for many decades America's coal heartland: Central Appalachia.

CHAPTER 2

KENTUCKY

The first time I met Danny Karst he asked me, "Where do your people come from?" Since I'm from Arkansas this was not a wholly unexpected greeting. "Little Rock," I told him. "My father grew up in Austin, Arkansas. Which is outside Beebe."

This seemed to satisfy him, but he had another important question. "What does your daddy do?"

"He's a retired insurance man." This, on the other hand, was a slight disappointment; Karst, a third-generation coal miner, reserves his highest respect for men who make their living with their hands, who've done manual labor for a big part of their lives. Preferably underground, like his own father.

I met Danny through friends in Kingsport, and he immediately agreed to spend a couple of days showing me around. At 48, Danny no longer mines himself; he's one of the lucky ones who got out and moved up in the world. His family—his three brothers, David, Dennis, and Darren, and his sister, Danielle—owns thousands of acres in Harlan County, Kentucky, once the richest coal land on earth. The Karsts don't operate mines anymore either; they lease the mines to the Harlan Cumberland Coal Company, a major force in Harlan County for decades. Danny and his siblings still make royalties from the mines (or, rather, they once again make royalties, having bought back the land after their father's company failed in the late 1980s). Harlan Cumberland is "the last operator still shipping coal out of Harlan County," Danny told me.

Founded by his father, Karst Coal Company thrived for three decades, in coal's heyday, until it fell victim to competition from Western coal, to tightening federal regulations on the coal industry, and, in Danny's telling, to the aggressive tactics of Karst's larger competitors. It rose again, with Danny and his siblings as owners/lessors, in the 2000s.

Today Danny Karst lives on a gracious street in Kingsport. He wore polo shirts and chinos the days I was with him, and drove a well-appointed SUV. He lives off of the proceeds from his father's mines, his own investments, and Karst Land, his development company, through which he has created a new Kingsport suburb called Edinburgh, not far from the neighborhood where he grew up. Talking to him you have the sense that he is slightly amazed by his own good fortune. And that he never forgets the toil and blood and hard history that went into making it.

In the story of Edward Karst, his mines, and his family can be read much of the twentieth-century history of mining in Harlan County, the heart of coal's Appalachian heartland. And in his son's life, in 2014, can be glimpsed the future of the coal industry in Appalachia, or what remains of it.

NO REGION IN AMERICA, and arguably no region on earth outside the coalfields of England and Wales, has been so identified with coal as eastern Kentucky. There is no economy here, except that of coal; there are no families whose histories have not been wrapped up in the coal industry; there are no politics outside the politics of coal. And there is no imaginable future that is not determined, either positively or negatively, by the future of coal. By the time I got to Harlan County, in the fall of 2013, the War on Coal and the decline of the coal industry seemed to be all that anyone thought or talked or wrote about in this part of the country. Even the prospects for the coming season of the Kentucky Wildcats, the state university's storied basketball program, seemed a distant concern as I drove the back roads and talked to the people and ate in the fast-food joints.

In fact, it struck me later that University of Kentucky basketball had evolved in a way that reflected the economic dynamics of the coal business. Under the roguish coach John Calipari, the Wildcats had come to specialize in a form of athletic carpetbaggery, hosting three or four top freshmen each year who were destined for the pros as soon as their nine months or so of indenture to the NCAA were up and they could apply for the NBA draft. It was successful, and it brought benefits to the state in terms of ticket sales and TV exposure and alumni enthusiasm, but it was a dispiriting way to run a program. The real wealth—the billions brought in each year by the league that treated the big state-funded universities like farm teams, and the million-dollar salaries earned by the mercenary athletes after they left Lexington—went elsewhere.

In the same way, the coal industry has kept Kentuckians fed and clothed and housed, but the big coal companies are headquartered elsewhere, the proceeds went mostly to wealthy executives and big shareholders, and the state, at least its eastern half, had been kept in a perpetual state of near-colonial dependency. Coal brought jobs but not prosperity, and as the jobs dwindled prosperity seemed more elusive than ever to the people of central Appalachia.

Coal was found here in 1750, a quarter century before the Declaration of Independence and four decades before Kentucky became a state. A remarkable physician/explorer named Thomas Walker, who crossed what is now the Cumberland Gap nearly 20 years before the arrival of Daniel Boone, built a bituminous campfire when he explored the region with the future Revolutionary War general Joseph Martin and their Indian guides. "They still had a little rum remaining," wrote Martin's son, "and they drank to the health of the Duke of Cumberland. This gave rise to the name of Cumberland Mountain and Cumberland River."[1]

Surface mining had begun by the early nineteenth century, and the first commercial mine, the MacLean Drift Bank, opened in 1820. By 1843 production of Kentucky coal reached 100,000 tons. The Civil War brought a significant boom to the mines of the region, but the postbellum era saw its first crash. Conflict has always benefited the coal industry in Kentucky—production hit 20 million tons in 1914, the first year of the Great War—but war's aftermath has always brought busts. In the post–World War II era, the mining companies mechanized, the railways shifted from coal to oil and diesel as their primary fuel, national environmental legislation was passed, and the center of America's coal industry shifted to Wyoming. Thus began Kentucky's long decline as a mining state. Coal mining employment had peaked in 1923, with more than 700,000 miners employed nationwide, perhaps 100,000 in Kentucky. The second half of the twentieth century also saw the rise of low-sulfur

eastern Kentucky coal, which came to dominate the industry as production in
the state's western counties fell, largely as a result of the Clean Air Act of 1970,
which for the first time placed restrictions on the emission of sulfur.[2]

And always, the region's dependence on coal brought a deep ambivalence.
The risk and relatively high pay of mining jobs bred "a pervasive fatalism"
among the miners, wrote Ben A. Franklin, a national correspondent for the
New York Times.[3] Mining was dirty and dangerous, but the pay was relatively
good and there were few other opportunities for the men who grew up in Appalachia. Management was cursed and derided by the miners, but woe to any
outsiders who questioned the reign of King Coal. The miner's dilemma was
that he was in thrall to the industry that damaged his health, exploited his
environment, and kept his community down.

Coal production in Kentucky peaked in 1990 at nearly 180 million tons,
but that year also brought the first in a series of laws that would effectively
drive many of the mines out of business.[4] The 1990 update of the Clean Air
Act was one of the most successful pieces of environmental legislation in history—it effectively eliminated the problem of acid rain in the United States—
but it was devastating for Kentucky's coal industry. Since the early 1990s the
charts tracing the state's production and mining employment trace a parallel
curve, steadily and inexorably down. Production from eastern Kentucky fell to
45 million tons in 2012 and continues to decline. The number of mining jobs
dipped below 20,000 in 1995 and stands at about 18,000 today.[5] And there
is no boom on the horizon this time. One-third of the 89 plants that bought
eastern Kentucky coal in 2012 and 2013 have announced plans to either retire
or switch to natural gas.[6] Many of the others will keep burning coal but won't
buy it from Appalachia anymore. Over the next two years, demand for eastern
Kentucky coal is likely to fall by another 35 percent.[7]

"We are in uncharted waters in southern and eastern Kentucky," said Representative Hal Rogers, announcing a December 2013 summit meeting on the
beleaguered region's economy. "The future of coal faces new regulatory challenges and economic uncertainty is daunting for our small communities."[8]

The list of social ills left behind by the coal industry's retreat stretches well
beyond daunting. Statewide, unemployment in early 2014 stood at around 13
percent, nearly double the national rate. In the eastern counties it was closer
to 20 percent. In Harlan County the number of coal mining jobs had fallen

by nearly half, according to the *Wall Street Journal;* in other counties, like Knott and Pike, the number of mining jobs wiped out was above 60 percent.[9] In 2012 and 2013 alone, more than 6,000 coal miners were laid off in eastern Kentucky. It was as if an economic bomb had detonated, leaving the miners and their families shell-shocked.

Many of these workers have no education beyond high school and few transferable skills. Accustomed to making good wages underground, they've taken on lifestyles, and accumulated debts, beyond anything they're likely to earn aboveground. And many former coal miners are unemployable thanks to the epidemic of prescription drug abuse that ravages many coal towns.

After years of crawling through dark tunnels to scrape coal out of the earth, miners tend to have plenty of bodily complaints. Many wind up on disability payments, and the physicians of the region have not been stingy in prescribing oxycodone, methadone, and Xanax. Kentucky is the fourth most-medicated state in the country, according to an analysis by *Forbes* magazine (coal mining states West Virginia and Tennessee are nos. 1 and 2, respectively).[10] More people died in 2013 from prescription drug overdoses than car accidents, and one of every three Kentuckians have a family member or close friend who abuses prescription drugs. The painkiller boom has produced its own black market, as unemployed addicts with prescriptions make money selling their extra pills on the side. Law enforcement officials in eastern Kentucky estimate that 80 percent of the crime in the region stems from the illicit drug trade, mostly prescription painkillers.[11]

"Prescription pill abuse is a reality that has touched the lives of almost every Kentucky family—including my own," Kentucky's attorney general, Jack Conway, told a hearing before the House Committee on Energy and Commerce in March 2012. "Prescription pill abuse has ravaged communities, shattered families and fueled crime."[12]

A group called Keep Kentucky Kids Safe, founded by two women from Morehead who lost their daughters to prescription drug overdoses, has formed to slow the drugs' spread. Committees, organizations, and conferences, in fact, are thick on the ground in eastern Kentucky, all of them trying to figure out what happens after coal. The most visible of these was the governor's summit, called Shaping Our Appalachian Region (SOAR), which brought 1,700 people, including most of the politicians and business leaders in eastern Kentucky, to the Eastern Kentucky Exposition Center in Pikeville for a day-long discussion of the area's economic future.

The meeting was earnest, well intentioned, and wide-ranging, but concrete results are hard to find. The main outcome of the SOAR summit may have been perceptual: it represented an acknowledgment that the coal era is ending, and something must be found to take its place.

"It's a different conversation this year than it was last year," Stephanie McSpirit, a sociologist and professor of Appalachian studies at Eastern Kentucky University, told me. "There's an opportunity now that wasn't there before. Now's the time that even the county magistrates, the county officials, and the chambers [of commerce] have realized we need to start investing and planning. People have started to ask, 'How can we think about promoting our region when we're destroying it?'"

That shift has only begun. Eastern Kentucky is still a place where people rally for the coal industry; where an ill-chosen remark in a bar can lead to trouble in a hurry; where, as I heard more than once from more than one person, "People around here don't like to speak against the coal companies." A popular bumper sticker around here reads, "Don't like coal? Don't use electricity." To raise money for its pro-coal campaign, Friends of Coal sells T-shirts that say on the front: "Legalize Coal." On the back is the familiar coiled snake and "Don't Tread on Me."

Treasury Secretary Jack Lew got a taste of the anger boiling out of Kentucky's coal towns in December 2013 when he testified before the House Financial Services Committee—not long after the Obama administration announced that the United States would no longer support funding from the World Bank and the Export-Import Bank for overseas coal projects.

"We want to see a future for the coal industry and the vitality of the American coal industry," Representative Andy Barr, a Republican who represents Kentucky's 6th congressional district, told Lew. "Kentucky's the third-largest coal-producing state in the country. And so my questions really are directed on behalf of the 6,200 coal miners in my state who have lost their jobs—with all respect—because of the anti-coal policies of this administration over the last couple of years."[13]

There followed the sort of tongue-lashing that executive branch officials are subject to when they must show themselves to hostile legislators on Capitol

Hill. Barr, in full spate, interrupted Lew several times to let him know that opposing new coal plants, in the Southeast or South Sudan, was tantamount to betraying hardworking Americans.

"As I tried to explain [earlier]," spluttered the harried secretary, "that we have very strong concerns about climate change, domestically and internationally, and we're trying to balance our concerns while promoting access to energy around—"

"I understand," broke in Barr, like a testy spouse. "I heard your testimony before."[14]

Speaking on the Senate floor a month later, Mitch McConnell, responding to the Environmental Protection Agency's proposed rule limiting greenhouse gas emissions from new power plants, echoed Barr: "Kentucky is facing a real crisis here. The Obama administration appears to be sending signals that its latest regulation is just the beginning in a new, expanded front in its war on coal."[15]

War looks different on the front lines than it does in the corridors of power, though. It's clear that there's a growing divide between the Kentucky congressional delegation, which represents the coal industry as much as it does the miners, and the mayors and city council members and county judges who have to look their neighbors in the eye every day. McConnell and Barr have the luxury of fulminating against the president and the EPA. The people of eastern Kentucky have more immediate concerns, like figuring out how to keep their refrigerators stocked and their houses out of foreclosure.

"I'm over the mad spell, blaming government, EPA," laid-off miner Garry Cox told a reporter from WFPL, the Louisville public radio station, at a job fair in Harlan. "Being mad, it ain't going to do nothing. It ain't going to help."[16]

You can shake your fist at the heavens, or at Washington, D.C., but ultimately you have to get on with life. Getting beyond anger, for the people of eastern Kentucky, is the next step in the grieving process that will free them from the past and allow them to start to figure out what comes after coal. Michael Cornett, of Harlan, runs a program called Hiring Our Miners Everyday, or HOME, based out of the Eastern Kentucky Concentrated Employment Program. Launched in early 2013 with a $5.2 million grant from the Labor Department, HOME tries to help laid-off miners find jobs in new fields. Part of the challenge is just convincing these men that the time has come to leave the coalfields. There are close to 8,000 unemployed miners in eastern Kentucky. As of the beginning of 2014, Cornett's organization had enrolled 1,400 of them. "The reality is, it's a difficult switch in this culture to get people to consider other alternatives," Cornett said.[17]

"WE BOUGHT THIS PROPERTY in 2005. The school opened in 2009. We almost lost all of it in the crash."

Danny Karst was driving me through Edinburgh, the housing development he and his brothers had built through his company, Karst Land. It was late afternoon, and bronze September sunlight divided broken clouds. Comfortable new homes, in a quasi-Craftsman style, dotted the lots around a pair of curving streets that surrounded the new elementary school. The 160-acre neighborhood, surrounded by forest on the southern edge of Kingsport, was barren of trees, for now; a fountain splashed into a new, two-level pond. Above the school a backhoe was parked at the end of one of the unfinished streets. Some of the houses were occupied, and some were for sale. The school, John Adams Elementary, runs not on coal but on geothermal energy, tapping the heat hundreds of feet below the ground. Danny Karst is a coal man to his marrow, but he was not biased when it came to his business. Geothermal energy made sense for the school and it made sense for the community.

Danny started building houses in Edinburgh in early 2007, just after the U.S. housing bubble peaked. "I felt like I needed to give something back to this area," he told me as we rolled slowly along the newly paved street. In 2007 Kingsport was going through its own housing boom, and Edinburgh was one of several new developments that sprang up in the area around that time, with names like Skyland Falls and Riverbend. The financial crisis, and the sharp downturn in the coal industry, has taken care of that boom; most of the other new developments have stalled or been postponed indefinitely. Karst Land has struggled on.

"We sold one house in 2010," Danny said. "We had to find a new market. I did the math and I figured out it would take me 187 years, at that rate, to pay off the loan. The help of our coal reserves is what has continued to help us charge ahead."

When I visited, Edinburgh had 17 homes, with another dozen or so under construction. It has been fueled by people moving to Kingsport to work at Eastman Chemical, which just opened a new 300,000-square-foot corporate business center, part of the company's $1.6 billion investment program called Project Inspire. Unlike the smaller coal towns that surround it, Kingsport is growing, becoming a regional medical hub in addition to the Eastman head-quarters, even as the coalfields that have historically supported the town die off. Danny Karst is among the lucky few who have made the transition from coal to different and more sustainable sources of wealth. "Today my family enjoys the benefits of the risks my daddy took. He always said the worst thing about royalties is it gives a sense of entitlement: the second and third genera-tions get lazy, and forget about how he went underground for all those years and made that money."

To say the least, Danny has not forgotten. The conversation about Ed-inburgh, and economic development in the Tri-Cities, led inevitability to the subject never far from his mind: the story of his family and of his father, Ed-ward Karst, who founded Karst Coal in 1960.

"In 1960 my dad started mining with $2,500 and a mule named Ellie. My grandfather spent 48 years underground and my dad was determined to own his own outfit. We've got drawings of his coal trucks from when he was five or six. He was a talented artist."

Danny showed me the drawings later. In a childish but competent hand, they showed big trucks with "Karst Coal Co." on the beds. The company would not exist for another 25 years. "He dreamed big," said Danny.

In the 1950s it was rare, if not unheard of, for a miner to scrape together the capital to buy his own land. Eddie Karst worked overtime for years, saving his money to start his own operation. He was 20 years old when he started mining on his own.

"Dad opened at Holmes Mill with a 28-inch seam. He sent a Polaroid to Tonawanda, New York, where they was needing coal. They bought Daddy's coal on his word."

The Niagara River town of Tonawanda needed Appalachian coal for its power plant, the Huntley Station, built in 1942, and for the coking ovens that had been built along the river in 1917. Eddie Karst had a live seam and a first customer for his coal. But he didn't have the funds to actually extract the coal and ship it to upstate New York. To get the coal moving he resorted to check-kiting, the 1950s equivalent of maxing out your credit cards. "His sister went around to banks all over Tennessee and opened accounts, and floated that money till he sold the coal and got it back. If they hadn't taken his word in Tonawanda that he could deliver that coal, he never would've gotten off the ground."

The coal delivered and the bankers satisfied, Eddie Karst's vision took physical form, complete with haul trucks that said "Karst Coal Co." on them. He eventually bought 300,000 acres of rich coal land in Harlan County, becoming one of the biggest independent producers in eastern Kentucky. The name Karst, of course, is aptly geological: much of the rock in central Kentucky is karst, or perforated limestone riddled with sinkholes, caves, and springs. Karst and coal don't always go together, but much of the coal in Central Appalachia is overlain with limestone karst. Danny's last name became associated with both the local geology and with his father, a coal man well known across eastern Kentucky who lived his life, and ruled his children's lives, according to bedrock virtues.

"We had a lawn service and he used to dictate what price we could charge widows. We could only charge 'em 50 cents to mow their lawn. Sometimes we only made a quarter. The gas cost a quarter, you know what I'm saying?

"He wouldn't give us anything. I got a job delivering papers, 'cause I was wantin' a cheeseburger. I saw Jimmy Trout eating a cheeseburger and I went home and said, 'Daddy could I get a cheeseburger?' He said, 'Yeah, if you get a job and you pay for it.'"

Danny was about nine. He did the math on how long it would take him to deliver enough papers to afford a cheeseburger and fries. Eddie Karst said, "If you take this job you have to keep it for a year." Ten months in Danny got sick. "They took me to the UK hospital. They thought I had leukemia, tell you the truth. Dad delivered the papers for me for a couple of weeks while I was sick, even though he was getting up at four o'clock every morning to go to the mines. I went to him, I was feeling pretty bad, and I said, 'Do you think I can quit now?'

"He was looking down at me and he said, 'Well, we have a reasonable excuse, so we're going to change our commitment. We'll change our commitment because of your illness.' He let me know how bad he hated to do that."

Eddie Karst was a church deacon in Kingsport, but his true faith was in himself and hard work and the sale price of coal. Theology captured his attention less than ethics.

"He heard me cursing one time. He never used explicatives. My brother and I were camping out in the yard and he heard me out there cursing. He asked if we enjoyed our campout, and we said yes, yes we did. He said, 'Well good, when's the next holiday?'

"Hell, we didn't know, we were eight years old or something. He said, 'You are confined to this yard till the next holiday.' It was Labor Day. Ninety days we were confined to the yard. He didn't cut it by a single day."

Karst Coal more or less scraped along through the 1960s but, like other mining companies in Central Appalachia, boomed in the 1970s as demand for Appalachian coal exploded after the OPEC oil embargo of 1973. The price of coal jumped 44 percent in one year. Coal exports from Appalachia more than doubled by the end of the decade, and while the biggest beneficiaries were the Big Coal companies like Arch Coal and Massey Energy, independent outfits with low overhead and rich seams (not to mention no unions), like Karst Coal, caught the wave as well. "The energy crisis," writes Ronald Eller in his masterful history of Appalachia since 1945, *Uneven Ground,* "temporarily restored faith in natural resource development as a private sector cure for Appalachia's problems."[18]

"Temporarily" turned out to mean a decade. By the early 1980s the center of the coal industry had begun to shift to the West, to Wyoming, where massive coal seams close to the surface enabled fully mechanized open-pit mining; a worldwide coal glut further reduced demand from Appalachia. The slump in coal prices coincided with two developments that would crush independent operators like Eddie Karst: the gradual decline of organized labor, under President Ronald Reagan, and the advent of new technology—particularly remotely operated longwall mining machines, which revolutionized underground mining. "By the early 1980s, larger, more heavily capitalized companies were adopting the new technologies," observes Eller. "Smaller operators, less able or willing to invest in the latest machinery, found it difficult to compete."[19]

As a child Danny watched his father's company materialize and prosper; in his twenties he watched it falter and die. Even as revenue declined and his machinery grew outdated, Eddie Karst stubbornly refused to sell out to a large conglomerate. His company slipped away from him in 1990, the year the updated Clean Air Act sent the Appalachian coal industry into a decline from which it has never fully recovered.

"In December 1990 we lost the operating unit," Danny told me. "Daddy didn't want a large company selling his coal—he was a stubborn man. I don't want to talk negative about anyone. . . . It was part our fault and part their fault."

"They" were his father's creditors. "They filed an involuntary bankruptcy against my father. The property was caught up in litigation for ten years after that."

Eddie Karst died seven years later, of a heart attack. He was 59. "I think it was the stress of working all those years," said Danny. "He spent 30 years caring for 80 or 90 families. He only had two layoffs in all those years."

Finally, in 2000, the bank put the former Karst Coal property up for auction. The land still had plenty of minable coal. The most likely bidder was Harlan Cumberland Coal, the operator that had worked the mines under a temporary deal with the banks since Karst Coal collapsed. Harlan Cumberland is owned by a contemporary of Eddie Karst named Joe Bennett. Danny had known Bennett since he was a boy. Danny and his brothers, with little capital of their own to bid with, decided to try and win back their father's land.

"We went to every bank in Kingsport. We found one bank out of 14 to lend us the money to buy it back from the trustees."

The auction was a scene repeated often over the years as the assets of failed miners were sold off to the survivors. Worth perhaps a few million dollars, Karst Coal was a minnow. Much larger fish would be caught up in bankruptcy court in the coming years.

"We went to the auction," Danny said, "and Joe Bennett was there to bid. We knew that he could've bought it 70 times over. There was no way we was gonna outbid him." Danny recited this part in practiced, almost verse-like cadences. He had told this story many times; it has become part of his family mythology. Only personal honor could resuscitate Karst Coal. "During the auction I took Joe out in the hall and said, 'Look, we've known each other all my life. Your dad went broke, my dad went broke. We have some commonalities. When is enough enough?'"

Bennett looked the younger man in the eye and said, "Danny, I just want to make sure that you'll give me the opportunity to lease it back."

"Yes we will," said Danny Karst.

Bennett held out his hand and said, "Danny, it's yours."

"He walked away," Danny told me as we left Edinburgh. "He plain walked away.

"Since then he's paid us 30 times over [in royalties] what he could've bought it for. And it's all based on a handshake. If it'd been me? I can't tell you I would've done that. That's the kind of man he is."

Coal continues to provide a handsome living for Danny and his family. "Our coal is expensive to get to but it's economic cause it's low sulfur, very low chloride, and it's hot. It's very hot."

"Hot," in this context, means a high energy content (as measured in BTUs) by volume. The coal from the old Karst coal mines is as high as 14,000 BTU per pound, compared to Wyoming coal, which is often less than 9,000 BTU per pound.

The creek-side coal seam that Eddie Karst discovered in the 1950s has funded the gracious homes of Edinburgh. Thanks to his father's stubbornness and his own persistence, Danny Karst is one of those who has not only survived but has prospered in the coal wars. He is not a man captive to his illusions, though, and he knows the future for Appalachian coal—his daddy's coal—is foreshortening rapidly.

"We produce about half a million tons a year. When we had the surface operations going it was about 100,000 tons a month. That was 2005 to 2008. So the underground portion has dropped from 45,000 to about 30,000 tons a month. We've lost about one-third of our royalties, mainly due to the market.

"The future's uncertain. There's nothing certain about it. I guess us coal people, as much as we get hit, we tend to be optimistic. The operator is looking at entering another seam that could be an opportunity. The question is, will we have a market to sell into?"

THE NEXT DAY WE DROVE NORTH, into eastern Kentucky. Danny's Uncle Bob, not a blood relation (and not his real name), drove the lead car and Danny and I followed in his SUV. We followed Highway 23 north into Virginia's southwestern toe to Big Stone Gap, where the three forks of the Powell River converge. Just south of the town of Appalachia, we climbed out of the Powell Valley and headed west, on Highway 68, crossing into Kentucky at Morris Gap. The mountains engulfed us. Danny's narration became a recital of ghost towns: this school closed in 2008; this family moved out, leaving the house intact; this fire station no longer had an engine. The road curled and swerved back on itself, befuddling my sense of direction. At one small hamlet we swerved to miss an old hound sunning itself on the asphalt; 100 yards later another lifted his head from his paws and regarded us with bleary disinterest. For some reason I thought of Argo, Odysseus's aged and blind dog, who recognized his master's voice after ten years' absence and then died.

"Still good coal and good people in these mountains," Danny said. "Now it's just the old folks staying here, the ones who can't afford to or don't want to leave. I don't know who's gonna be left here when they're gone."

The depopulation of Central Appalachia, in progress to some degree for decades, has accelerated since the turn of the twenty-first century. On the list of challenges facing America, the decline of a region that has always been poor, rugged, and thinly populated doesn't rank very high. In strictly economic terms, it's rational for the young people of eastern Kentucky—at least those who can—to get educated and get out. Atlanta, Pittsburgh, Knoxville, Lexington, and even the smaller cities of Kingsport and Greensboro profit from the influx of labor and human talent. Indeed, a school of economic thought arose in the 1950s, a time when most of the country was undergoing a postwar economic boom, that asserted the answer to Appalachians' misery was not local development but diaspora. "Economic betterment for the great majority of the people of the Southern Appalachians is not to be found in development of the meager resources of the local area," wrote economists B. H. Luebke and John Fraser Hart in 1958, "but in migration to other areas more richly endowed by nature and by man."[20]

People should be free to go where there's work, and schools, and a future. This is America. But the quasi-feudal economy and the deep ties of family and community in coal country make moving on a lot more complicated than the simplistic models of economists. Coal people tend to stick and stay, and resist change. And forcing them to do otherwise is a craven way of wrestling with the forces at work in Appalachia.

Telling the people of Holmes Mill and Big Stone Gap and Hazard "just move to the city" not only ignores the ties to the land that knit a society together, but it's also bad policy. The health of a society, and of a nation, can be measured in the well-being of the least fortunate. And out-of-work coal miners in Harlan County have moved ever closer to the top of that list.

The United States has done a very poor job in recent decades in supporting, preserving, and rejuvenating communities left behind by globalization and other broad economic shifts. Letting Detroit go bankrupt, letting steel towns like Bethlehem rust away, letting once prosperous farm towns in the Mississippi Delta go to seed is not just unfair; it's unwise, as crime rates, public health statistics, and welfare rolls all eloquently attest. Leaving the coal towns of Harlan County to wither says something discouraging about the American project. The tides that have washed over and receded from eastern Kentucky will reach all of us soon enough.

"Appalachia's problems are not those of Appalachia alone," writes Ron Eller in *Uneven Ground*. "We are all Appalachians."[21]

Gradually, tortuously, the road crested and dropped and we came into the deep chasm of the Clover Fork of the Cumberland River, which roughly parallels the Kentucky-Virginia border and which brought us to Holmes Mill, the village where Eddie Karst staked his first mining claim. We pulled off by the post office, abandoned within the last year or two, and Danny pushed aside the streambed underbrush to show me the seam where his father first found coal. A few rusted rails still clung to the side of the creek where a belt-driven conveyor took coal to the tipple, or loader. Danny stared at it awhile.

"That there was the fan house," where a big fan forced fresh air into the mine for the miners to breathe. "My daddy had a tipple right up there, he leased this land. He was swimming in a hole up the river and he saw this upcrop. That would've been 1959. He was 19. That's when he decided he was going to mine this land. That's where he got that dream."

After two days of traveling with Danny, I'd started to hear a compulsory quality to his storytelling. He seemed like a storm-tossed mariner who'd found himself stranded on a strange and rocky shore, forced to keep telling this story in order to make sense of it. We walked back across the road. A couple of girls in their teens or early 20s were walking a slow-moving black lab along the abandoned railroad tracks that led behind the row houses that formed what remained of Holmes Mill. Built by the coal company in the 1920s, the houses were of the type that can be found across Appalachia: wood-framed, narrow, two-storied, with a shotgun hallway on the ground floor. The only differences from the Depression were that there were fewer people living in them and several had big, tricked-out pickup trucks parked outside. It's a good time to be in the vehicle repossession business in eastern Kentucky.

We drove a quarter mile up the road and parked by a small patch of lawn with a pavilion, a basketball court, a stone fountain built into the hillside, and a couple of picnic tables. Harlan Cumberland Coal owns the park but Danny Karst pays for the upkeep. "They don't have the money to maintain it," Danny said. "I wanted it to look just like Edinburgh. I thought it was time to give something back to this area."

Danny, Uncle Bob, and I stood listening to the insects chorus in the woods. A worker on a riding lawnmower pulled up next to the pavilion.

"You don't remember me, do you?" he greeted Danny.

"You're Milt's boy, ain't you?"

"That's right, Milt Saylor. I'm Milt Jr. We lived up 'ere on Black Mountain. I remember y'all lived up north of us."

"Yeah, yeah—I went to school with your brother. How old are you?"

"I'll be 50 on my birthday."

A haul truck with a dark blue bed, unmarked, rumbled down from the ridge above us with a load of coal and onto the road. It disappeared to the west.

"David lives down there in Harlan," said Danny. "You remember Dennis, he was the mean one. You got kids?"

"I got two boys, one still at the house. I can't run him off. The other's a machinist up in Whitesburg. Makes bearins and things. He can make anything. Y'all still mining coal?"

"Joe's mining it. He leases our land."

"Your daddy was sho a fine man. He was always good to us kids. You don't forget stuff like that."

"Y'all had the store, didn't you?"

"Yep. We had Saylor's grocery."

Another pickup pulled up and out of it unfolded a tall man in jeans and work boots and a canvas work shirt worn smooth to the consistency of linen.

"This here's Roy Bush. He was one of Dad's favorites. Dad used to say, 'If you could work like ole Roy you might be something.'"

Roy shook my hand and smiled out of a face as seamed as the surrounding mountains. If Jimmy Stewart had ever starred in a movie about an aging Kentucky coal miner, he would have played Roy Bush. We stood chatting a while in the warming sun. Danny went off with Milt Jr. to discuss the state of the park. I asked Roy about his life.

"I was born there about half mile up the holler. I worked for Mr. Karst from 19-and-77 till 1991, when they done shut down. You know how kids have their plans, what they want to do? I wanted to go to work for Eddie Karst when I was 18. And that's what I done."

Roy worked as a common laborer, at first, and then a shift boss. He became a trusted lieutenant to Eddie Karst and an uncle figure to Danny and his siblings. Like many of his contemporaries, he made a life in the mines that was, by eastern Kentucky standards, enviable.

"I made $5 an hour. That was a lot of money back then. In 1991 I was making $18 an hour an' I thought I was rich. When Danny gets back over here I'm gonna tell him how much cheaper he is than his dad."

Now he was semi-employed. He did some occasional jobs for Danny at Edinburgh. He sold ginseng that ex–coal men even more hard up than him collected along the hillsides and the canyons—ginseng collection and sale is a rare growth industry in Central Appalachia. He found ways to get by. He had a cabin up on the ridgeline above the mines, where he and his wife plan to retire one day. For now, like Bill Pritchard and many of the employees of TVA and many, many lifelong coal miners, he was too old to start over and too young to die.

After a while, having properly appreciated the pleasures of the park, we got back in the cars and drove up to the Harlan Cumberland mine, which clung to the north canyon-side just above Holmes Mill. A sign said "Karst Robbins Coal Co." Halfway up the steep access road we pulled over to let a coal truck get past. At the top there was a clearing with a broken-down conveyor belt line and some rusting metal containers. "My daddy designed that on a paper sack from scratch," Danny said. "He had it in his head and finally got it down."

Beyond was the working mine, looking not much different than it probably had 30 years before. From a cavern in the side of the cliff face ran a belt line, rattling continuously, that occasionally belched out a few hundred pounds of coal ore. From the pile underneath the conveyor, a Caterpillar loaded the coal into trucks that showed up every few minutes and then headed off back down the hill. A pair of metal stronghold doors gave access to the mine. A few trucks were parked in the lot next to the shack that served as the operations center. The whole operation had a low-tech, pre–Information Age look. Carl, the mine boss, who declined to give me his last name, showed me around. He was big and boisterous and unswerving in his views on the War on Coal—if Roy Bush was Jimmy Stewart, Carl was Fess Parker in his fleshier, post–Daniel Boone days.

"This here's part of the Darby Seam," said Carl. "It's one-point-five mile deep in there. Them seams are layered, that one you're lookin' at is a bastard seam.

"They use a ripper head at the coal face, to cut the coal. It's all remote control. They's probably 25 men in there right now. They get down there on mantrips, we call 'em buggies."

"Does this mine make money?" I asked, innocently.

"Barely. Over the life of the mine, it takes about $5 million in supplies ever year since 2008. The belts are $50 a foot."

"Installed?"

"Hell naw, that ain't installed, that's just to buy it. We got 12 miles of belt in the mine. The structure's $14 a foot. The high lines, the high-voltage cable, that's $13,000 right there. You're looking at $200 million worth of shit here. And we don't make none of it. Every layoff in the mines means 17 others lost their jobs too."

He showed me a steel drill bit, swirled and pointed and beautiful in its way, like a warhead. "There's 70 bits per ripper head. We change out a dozen bits every cut. I buy two pallets of bits a week. That's about $9,600 on bits alone per week for the face. Another $8,000 in roof bits per week to drill the holes for the supports.

"You can't make it at $50 a ton anymore. When I was young you wished you could get $50 a ton. You know what it is? Regulations."

This was Carl's true theme, and he went at it with gusto for the next half hour or so. A couple of times Danny elbowed me in the ribs and murmured, "You don't need to write that," so what follows is edited to remove some of the more inflammatory parts of Carl's rhetoric. Mine inspectors came in for his especial scorn. Inspectors, he said, visited the site daily to demonstrate their industriousness. There were too many of them, and the state had not seen fit to reduce their numbers. "In 2010 they was 300 mines in Harlan County. You know how many there are now? Less than ten. Less than ten.

"They're talking about disbanding the state inspectors, layin' 'em off. They got nothing for 'em to do. It's bullshit."

This despite the 2006 explosion and fire at the Darby Mine No. 1, just up the road, that killed five men (which Carl first mentioned to me). "The accident occurred because the operator did not observe basic mine safety practices and because critical safety standards were violated," the U.S. Mine Safety and Health Administration's investigation concluded.[22] It's hard to observe all the safety standards when the mines are unprofitable and men are desperate for jobs. The year 2006 was an unusually deadly year: 15 men lost their lives in coal mine accidents in Kentucky. Since then the fatality rate has been steady: about five per year, enough to devastate a bunch of families annually but not enough to make national news. For that you need a full-on disaster like the Upper Big Branch explosion in West Virginia, which took the lives of 29 men in April 2010.

Along with Carl and three other miners, Danny and I crowded into the operations shed. On one desk strewn with papers was a Mine Boss fire prevention monitor; another monitor showed the closed-circuit video of the face where the men were working. On the wall a grid map showed the mine divided into squares, some shaded with lines, some blank, and some X'd out. An X meant that section couldn't be mined, per the regulations. "We can't mine it because some guy in Pittsburgh with a computer model says the mine'll collapse if we do," Danny said.

The Harlan-Cumberland is a longwall mine, which means the mine face is cut into a long slice or wall about a yard thick and a few hundred yards wide. The coal along the face is cut away and the overlying rock falls into the corridor behind it. Developed in the late seventeenth century in England, this method was originally known as the Shropshire method. The vast open-pit mines of Wyoming, China, and Mongolia—not to mention the mountaintop removal mines in West Virginia—have made this a dying art. Coal that has to be gotten from a longwall mine, deep underground, in the twenty-first century, is seldom profitable.

"We can't mine but one-third of that panel," Danny said, stabbing a finger at the grid. "We got 80-by-80 square pillars in there to support the mine. It's a retreat mine, which means you take it out as you back out of the mine, and you actually want it to collapse as you retreat 'cause it relieves the pressure."

"It's a good honest profession," Carl said, looking at the floor and shaking his head. "They're trying to end it in this country."

In the shed it was stuffy and mournful, like a wake for a relative who hasn't quite passed yet.

"These guys, most of 'em are going to leave," Danny said, indicating the men in stained overalls leaning against the walls of the shed. "You can't replace that knowledge, you know what I'm sayin'? Once you lose them, that knowledge is gone. It's not like natural gas where you just come in and drill and frack it out and move on. You've got to have that experience and expertise of men who've been in the mine and know what they're doing."

"So how much longer do you think you'll mine here?"

"A year ago I'd've said eight to ten more years. But now, with the new modeling they're using, it's probably down to six. Then it'll be closed forever."

AND WHAT, THEN, WILL BE LOST? The livelihoods for a few dozen men, certainly; a portion of Danny Karst's family's wealth; another shred of the tattered mining legacy of eastern Kentucky. A fragment of American history, as they say, in a dark and bloody place. A few hundred megawatt-hours of cheap electricity that will be readily replaced with cleaner sources, and a few thousand tons of carbon dioxide released into the atmosphere that will, hopefully, be eliminated.

That didn't ease the despair and frustration of the men at the Harlan-Cumberland mine. It's a hard thing to watch the business that has supported your family for generations slipping quickly into history. Six years, eight years, maybe ten. The whole region was just hanging on, clinging to an industry and a way of life that had brought neither dignity nor prosperity to those born into it but whose passing, without a credible substitute, will leave holes as large as the underground mines that pock these hills.

The futility was compounded by the realization that the broader coal industry, like a dowager enjoying a fleeting springtime, was experiencing something of a comeback. Demand was climbing again, and Wall Street analysts had begun to talk of a "bottom" to coal company shares; the historic low natural gas prices of 2011–12 had proved unsustainable, and the harsh winter of 2013–14 was about to push coal prices upward again. Like the Red Army after Stalingrad, the coal business was recovering. Just not in Harlan County.

You could find signs of the recovery just on the other side of Kentucky, in McLean County, where a company called Rhino Resource Partners was about to open the Pennyrile Mine, the first new mine in Kentucky in a decade. Chris Walton, Rhino's new CEO, said that Rhino would double its production at Pennyrile from 800,000 tons a year at the outset to nearly 2 million within a few years, bringing in $20 million a year. The main reason eastern Kentucky was foundering while new production was going on a few hundred miles away is that Kentucky sits atop a great geological divide that separates the Illinois Basin in the west from the Central Appalachian basin in the east. Illinois Basin coal is high in sulfur but relatively cheap to mine, and unlike places like Holmes Mill, the best seams have not been scraped out. In one of the ironies that beset the industry in the twenty-first century, stricter air pollution regulations had forced most utilities, by 2013, to install scrubbers on most of their plants, depleting the value of low-sulfur Appalachian coal. Rhino already had a contract to supply 800,000 tons a year to an unspecified utility through 2017. Long-term contracts like that were scarce in Harlan County.

Rhino is based in Lexington, a gracious university town in the heart of bluegrass country with a faint but persistent odor of old money and fresh horseshit. Although Frankfort is the state capital, Lexington and Louisville are the true centers of wealth and power in Kentucky. Lexington is less than 150 miles from Holmes Mill, but it's so distant in socioeconomic terms it might as well be Budapest. It's also the hometown of Alison Lundergan Grimes, the young, attractive, and dynamic secretary of state. Grimes, who graduated from

Lexington Catholic High in 1996 and was challenging McConnell, the powerful minority leader of the Senate, in a race that CNN said "could be the most expensive and nasty 2014 Senate race in the country."

Like Hillary Rodham Clinton once had, Grimes made a point of her middle name, not to distance herself from a philandering husband but to remind voters of her father, Jerry Lundergan, the longtime chair of the Democratic Party in Kentucky. Like all candidates for statewide office in Kentucky, Grimes was vocally pro-coal. She promised to break with the Obama administration over its aggressive regulatory assault on the industry. That didn't stop McConnell and his supporters from insinuating otherwise.

"Alison Lundergan Grimes stood proudly at the Democratic National Convention to nominate Barack Obama, who is now following through on his 2008 promise to destroy the coal industry in essence declaring a war on the state of Kentucky and the middle class families who call it home," declared the National Republican Senatorial Committee, in a campaign email that went out a few weeks before I visited Harlan County. Bill Bissett, the president of the Kentucky Coal Association, went further, warning Grimes that she'd "better not" accept campaign donations from Tom Steyer, the California billionaire and former hedge fund director who was pouring his considerable fortune into the struggle against climate change, including the anti-coal movement. Grimes had reportedly attended a meeting where Steyer was present. She denied meeting him and pledged to help roll back federal restrictions on coal plant emissions. As is often the case, Kentucky politics were largely a contest to see who could most passionately support the coal industry. "You can either be pro-coal or accept money from [Steyer's] political network, but you can't do both," declared Bissett.

Guided by the United Mine Workers, Appalachians for generations voted Democratic, but Barack Obama and the so-called War on Coal had turned most of the region deep red. Nevertheless, by summer Grimes had managed to keep pace in raising money with McConnell, who has received hundreds of thousands of dollars from the coal industry over the years. The Kentucky Coal Association, the coal companies, and the national Republican Party were pouring their might into returning McConnell to D.C. ("Obamacare. The War on Coal. That's Obama's agenda. And Alison Grimes supports Obama," ran one TV ad sponsored by a pro-McConnell super PAC.)

After five terms in Washington, support for McConnell was hardly passionate in Kentucky. But Grimes could never quite convince the voters she was truly a friend of coal. McConnell beat her handily and became the new Senate majority leader. Coal still has powerful friends in Congress.

"I'm pretty tired of the coal corporations and coal barons telling our elected leaders what they can and cannot do," wrote Carl Shoupe, a former coal miner who was almost killed in a mining accident that ended his career underground, in an op-ed in the *Lexington Herald-Leader.* "But I am absolutely sick and tired of political leaders—or candidates—who let them. We are starved for leaders who will look out for Eastern Kentucky instead of doing what the coal companies tell them. We are ready for leaders who will help us build the bright future we deserve here."[23]

That future was hard to see as Danny and I, followed by Roy Bush in his pickup, drove up a dirt road that followed a wide slash of bare ground along the mountainside above the Clover Fork. The open space was a former strip mine, abandoned three years before. Danny wanted to show it to me to counter any notions I might have about the destruction of the environment. The hollow, shaped like a closed amphitheater spilling out onto the slope below, was grassed over. On one side was a long sluiceway filled with boulders. A few willow shrubs and pine saplings had started to recolonize the open ground.

"See how this looks?" asked Danny. "It looks pretty good, don't it?"

I mm-hmmed noncommittally. We pulled over where the road flattened out, next to a pile of rubble and a broken-down backhoe. Derelict machinery is the signature art form of the Appalachian coal industry. An empire had passed, leaving behind these rusted implements. The backhoe, Roy Bush said, had been stripped of its usable parts only days before by a thief who'd been desperate enough to come all the way up here to steal them. In World War II, on the Eastern Front, Russian partisans stripped Panzers abandoned by the retreating Wehrmacht. Here in eastern Kentucky, the inhabitants were cannibalizing the remnants of their own industry. Roy shook his head, considering the deeds of men.

"I thought about calling the police," he said. "Then I went home and thought about it and I said, 'How close am I to that? To havin' to do something like that. Do I want to put that ole boy in jail?'"

Danny, as usual, found a parable in the situation. "I asked in Bible study class a few weeks ago, 'Would you go in another man's field and steal the crop from his fields?' Like the Bible says. They said, 'Aw no, we wouldn't do that.' I said, 'That's bullshit.' You don't know. You don't know what you would do when it comes down to feeding your family. No one does."

"That's what it's comin' down to in Harlan County," said Roy Bush. "That's what it's comin' to."

CHAPTER 3

WEST VIRGINIA

I said goodbye to Danny Karst and Roy Bush and continued on Highway 38, winding up the Clover Fork Valley and crossing the ridge and then down a series of tortuous switchbacks to Harlan. I turned east, on U.S. 119, which follows the Poor Fork of the Cumberland across Kentucky. A few miles up the road I passed Harlan County High School, opened in 2008 to replace three abandoned high schools in the region. Like many kids in the region, Roy Bush's high school–aged son travels 25 miles by bus through these hills to get to school every day.

Linking the mining towns of the Cumberland Valley, 119 is one of the coal thoroughfares of Appalachia. Every few miles I saw an abandoned tipple, a riverside load-out, or an equipment graveyard. At Cumberland, a short jog to the east on Highway 160 takes you to the hamlet of Benham, home of the Kentucky Coal Mining Museum. Just past Cumberland the road begins another series of switchbacks into the George Washington National Forest, climbing past the headwaters of the Cumberland. Bending north, I passed out of the Cumberland Valley, through Pikeville and across the state line, marked by the Tug Fork River, into Williamson, West Virginia.

Once a hub of the Appalachian coal trade, Williamson has a large rail yard, built in the early twentieth century by what was then the Norfolk and Western Railway, and once had a series of landing stages along the Tug Fork where coal was loaded onto barges for the trip north to the cities of Pennsylvania and Ohio, even Detroit. The town boomed after World War I, reaching nearly 10,000 people on the eve of the Great Depression. Today, the population is less than 3,100 (the town lost nearly 10 percent of its inhabitants between the 2000 and 2010 censuses), and Williamson, like many once-thriving cities across Appalachia, is struggling to survive.

That struggle, as always, has attracted a steady stream of studies, books, position papers, and especially speeches. Many, many politicians have spoken about how they are going to save West Virginia, turn around the coal industry, and bring modernity and prosperity to Appalachia.

"If new vision, and new leadership, and new understanding can restore prosperity to the coal industry—and I believe it can—then I am here to tell you that prosperity is on the way," said one such orator. "America has a rapidly expanding population and a growing economy—more Americans and more industry will need ever-increasing supplies of fuel and power—and coal is our most abundant and economical source of energy for the future. . . . Perhaps even more important is the increasing productivity of the coal industry—and the important new markets, new uses, and new applications for coal which are within the reach of our science and our technology. . . . And we must act now to meet this challenge if coal is to be in the future—as it has been in the past—the foundation of American strength and the source of American plenty."[1]

That was John F. Kennedy, on a campaign stop in Morgantown in April 1960. Kennedy came to Appalachia because, as a Catholic senator from Massachusetts, he needed to prove that he could win a rural, overwhelmingly Protestant state—and his primary victory in West Virginia would propel him to the Democratic nomination for president. He would later credit his success in West Virginia for landing him in the White House. Political considerations aside, Kennedy, our most empathetic president until the emergence of Bill Clinton in the 1990s, was genuinely appalled by the conditions he found in the coalfields. A tireless campaigner at 43, he spoke at coal camps and town halls and large auditoriums. He called West Virginia "an island of poverty and distress in a vast sea of American plenty."[2] And he promised the West Virginians a brighter future, based entirely on the bright prospects for coal.

"We must establish a national fuels policy," Kennedy declared, "greatly increase our coal research, and stimulate industrial development. But perhaps the most visionary, and yet the most promising and the most fruitful prospect for the future of coal lies in the development of new steam plants—in the increased use of what can be called 'coal by wire.'"[3]

Coal by wire meant, simply, mine-mouth plants: burning the coal at the source and building expansive transmission systems to the north and east. Four major cities—Chicago, Detroit, Philadelphia, and New York—lay within a 500-mile radius of where Kennedy stood. Creating great steam plants near the coalfields would cut down on transportation costs and create "vast new markets" for West Virginian coal.[4] Kennedy predicted that average electricity use would jump from 3,300 kilowatt-hours (kWh) of electricity per household, the equivalent of about a ton and a half of Central Appalachian coal, to more than 22,000 kWh (11 tons of coal) within the decade. Coal now, coal forever: "the foundation of American strength."[5]

Kennedy's forecast for electricity demand was far off—by 2012, the average American household used just 1,837 kWh—and his vision for the future of West Virginia was even further off. Coal by wire never materialized; transporting coal out of towns like Williamson was cheaper than building big plants, and the accompanying transmission lines, in rugged West Virginia. Politicians and coal industry executives are still talking about the fabled "new markets and new applications" that will, one day, materialize. Fifty years later, conditions in the coalfields have improved—very few people live in dirt-floored shacks, and even fewer go hungry—but West Virginia remains a backward and downtrodden state, an island of want in a sea of plenty. And it was a very different kind of speech that another politician with a famous American name gave recently on the floor of the U.S. Senate.

The date was June 20, 2012; the occasion was one of the minor, procedural votes that clog the halls of Congress on slow news days in summer. James Inhofe, the climate change–denying senator from Oklahoma, had introduced a resolution of disapproval of the Mercury and Air Toxic Standards, or MATS, the Environmental Protection Agency's new rules on mercury and other forms of air pollution from power plants. Jay Rockefeller, the five-term Democratic senator from West Virginia and the great grandson of John D. Rockefeller, rose to speak against the Inhofe resolution.

"I rise today in the shadow of one seemingly narrow Senate vote . . . to talk about West Virginia," said Rockefeller. "About our people—our way of life, our health, our state's economic opportunity—and about our future."[6] That future, Rockefeller declared, lay not in clinging to the past, to the old ways and the old economy. The scion of America's fossil fuels dynasty was for the first time speaking against coal, and against coal's grip on West Virginia.

"Coal has played an important part in our past and can play an important role in our future but it will only happen if we face reality," Rockefeller said. He went on to excoriate coal company executives who want to "somehow turn back the clock, ignore the present and block the future," who "would rather attack false enemies and deny real problems than find solutions." They are "abrogating their responsibilities to lead," Rockefeller charged.[7]

"Stop and listen to West Virginians—miners and their families in-cluded—who see that the bitterness of the fight has taken on more importance than any potential solutions. Those same miners care deeply about their chil-dren's health and the streams and mountains of West Virginia. They know we can't keep to the same path."[8]

Rockefeller didn't offer many specifics on an alternative path—he called for "a smart action plan that will help with job transition opportunities, spark-ing new manufacturing and exploring the next generation of technology." It was like calling for another report, another blue-ribbon committee to examine the future and come up with recommendations that would mean little to un-employed coal miners. And Rockefeller himself had little to lose: less than a year later he would announce his retirement from the Senate. But in declaring that it was time to face facts, in specifically calling on the coal industry to "dis-card the scare tactics" and "stop denying science," he became the first national politician from a coal state to openly break with the industry. And he put a new frame around the fight over West Virginia's future.

LIKE THE HILL-COUNTRY FEUDS that rived West Virginia in earlier genera-tions, the coal wars have pitted town against town, neighbor against neighbor, family against family. Kentucky has the bluegrass country, the Wildcats, and, in Louisville, a thriving, cosmopolitan Ohio River city of a quarter-million people. West Virginia has coal. To be sure, the dependence is largely psycho-logical: coal mining actually contributed less than 7 percent of West Virgin-ia's GDP in 2013. Today fewer than 23,000 people work in the coal industry in the state, out of a total workforce of more than 770,000.[9] But those are high-paying jobs; coal miners make an average of $68,500 a year, according to the West Virginia Coal Association, about twice the state average. Coal companies pay 60 percent of the business taxes in West Virginia, around $600 million a year, a figure that is declining annually as production falls.[10]

And it's dangerous to suggest publicly that coal is not the future. After becoming nationally recognized for her struggle against mountaintop removal mining, activist Maria Gunnoe, of Boone County, began seeing her face on "Wanted" posters around town. In 2012, when she testified before a congressional hearing on "Obama's War on Coal Jobs" about the damage coal mining has wrought on the countryside where she lives, Republican representatives accused her of possessing child pornography after she attempted to display a photo of a five-year-old girl bathing in tea-brown water. She was detained and questioned—and later released without charges—by the Capitol Police.[11] It was among the most egregious examples of coal's supporters trying to silence opponents, but it was hardly unique. The harassment of people who express less than fully enthusiastic views on the industry seldom becomes actually violent, but it is real.

"I've got 40 acres of land that's been in my family since 1951, and I've got 400 feet of chain-link around it with four German shepherds," Gunnoe, whose father was a coal miner till his death at age 51, and whose son is a miner now, told me. "They've threatened to kill me, and burn my house down. I've had coal trucks barreling ass at me trying to run me off the road.

"My grandfather built those roads and I refuse to be run off 'em. We are winning this fight, and it risks my life every day."

I talked to Gunnoe the week after the new EPA rules on power plant emissions were announced, an event that was as momentous to West Virginians as the Emancipation Proclamation was to southerners.

"The only real question is where on a scale from devastating to a death blow the new rule will fall," said Alex Mooney, a Republican candidate for Congress. "I, and I am sure others who represent coal mining communities, will not sit idle in the face of this latest challenge by EPA to our way of life."[12]

The coal wars make people think and speak in biblical terms. Environmental catastrophe on the one hand, economic meltdown on the other. Since at least the 1930s, when Governor Herman Kump, who served from 1933 to 1937, railed against FDR's New Deal and uttered dark warnings of "federal subordinates" who were undermining the Appalachian way of life, West Virginians have drawn on a deep well of mistrust of Washington. Moonshiners, miners, ministers, and mayors in West Virginia have been railing against the feds for more than a hundred years, and today the struggle to preserve the coal industry takes the place of realistic efforts to devise a new roadmap to prosperity in Central Appalachia.

"It's a great tragedy what's happening here," Ted Boettner, the executive director of the West Virginia Center on Budget and Policy, told me when I met him in a Charleston bookstore/coffee shop. "Instead of putting all this effort into fighting the EPA, it's critical that we put our energy into creating policies that will give us a chance to create a new future for people in this state."

A MAN STANDS ON A HILLTOP in southern West Virginia and points to the horizon. The mountain has been scraped bare for a coal mine. The surface is dirt and crushed rock, compacted. It looks like Afghanistan, only flat. There's still some mining going on in one corner of the property. There's also a hardwood flooring factory and a concrete plant and a parts supplier for big mining trucks. The former mine land is being converted, gradually, to an industrial park.

"Right over there is hopefully my mountain," he says, pointing to a verdant knoll to the southeast.

The man's name is Eric Mathis, and though he's only 37 he's traveled a long way to get here. He talks in a rush. "Right over here is where the solar panels will go." He scratches a rough map in the dirt. A solar farm, a natural gas–fueled power plant, a hemp processing plant. Groves of sycamore and black locust. New jobs in sustainable, locally owned businesses. The barren ground takes a new and fertile shape in his imagination.

"Here's your biodiversity, here's your carbon sink. OK, locals, here's your economic diversification. We're trying to build that middle bridge, to involve everyone who has a stake here. Here's your new industry. The wood-flooring plant that's already here employs 150 people. Here's where the building pad for the natural gas facility will go."

We're standing on the James H. ("Buck") Harless Wood Products Industrial Park, about 30 miles outside of Williamson. All told it's 650 acres, with power and water lines installed and broadband Internet access established. It was designated an industrial park on reclaimed mine land in 1999; there are plenty of sites still available. To Eric Mathis, this is what the post-coal future in Appalachia could look like. "The idea is to bring industry in that can take advantage of what we have right here. The problem with economic diversification in Appalachia is there's a lack of spin-off industries—in an extraction economy they come in and extract the resources and run off somewhere else and upgrade them. What we're doing here is extracting the timber resources and upgrading them right here.

"The same thing with the coal: we're extracting and upgrading it with the concrete facility. It's adding value that returns to the local community. It's something that's never been done before in Appalachia."

It hasn't really been done here yet, either. The wind sweeps across the Harless Industrial Park. Dirt roads crisscross the site, with little traffic. The plans for an economic renaissance are mostly theoretical, for now. There's infrastructure, and there's 250 jobs or so, but not much industry yet. The man-made plateau on which the park is located is three and a half miles up a steep road from U.S. 119. There's no railroad, no river port, and no freeway within 70 miles. We are surrounded by forest, rivers, and hollows, 80 miles from Charleston, 160 miles from Lexington, 520 miles from Cincinnati. The nearest airport is 60 miles away.

"The enviro community says nothing will ever grow back here in a million years," Mathis says. "That's bullshit. There's been six glaciations through here and the forest always regenerates. If you think mountaintop removal is bad, think of this entire region after the glaciers came through."

Mathis indicates a green knoll a few miles off, beyond the edge of the mesa. It has an artificial, manicured look that lacks the wildness and chaotic vegetation of the surrounding mountains. "I call that a poor excuse for a mountain. You can restore the land to something that resembles how you think it was before it was mined," he says. "Or you can use the land so it benefits the community and the people that live here. The other way doesn't provide any valuable asset. This does."

Mathis's vision tumbles out so rapidly that it's hard to follow at times. Give him a crowd and a podium and he could be a junior senator from Massachusetts, circa 1961. Power purchase agreements, tax offsets, reclamation permits. A 69-kilovolt line running up to the site. If you squint hard you can just see the solar panels in long lines, the power plant, the new factories, the workers arriving for their shifts. In West Virginia, the future is a long time in coming.

ERIC MATHIS GREW UP in the less desirable environs of the tobacco capital of Winston-Salem, North Carolina. Part of his childhood he lived in the Peace Haven trailer park. He was not a promising student. "I was never encouraged to make anything of myself," he told me. He dropped out of high school, did serious amounts of dangerous substances, "went down the route of white trash." His life has been full of detours, false starts, and diversions. He got clean, went back and got his adult high school diploma and spent some time at the Art Institute of Atlanta, but "quickly realized that I was being funneled into a cubicle Nike corporate job." He dropped out again.

He began to find a direction when he fell in love with climbing and philosophy, which brought him back to the Appalachians, to Boone, North Carolina, where he worked his way through community and technical college and eventually enrolled at Appalachian State University. The terrorist attacks of 9/11 caused him to once again reevaluate his worldview, to "question what America is all about," and he dropped out again and became an executive chef at a North Carolina resort. He loved it, but that didn't last either. He is one of those people who cannot live a life without a larger meaning. He went back to school, enrolled in Appalachian State's interdisciplinary studies program, started showing up at local poetry slams, got involved in the community garden movement, raised money to help immigrant farm workers organize, and discovered that he had a gift for community activism. "Mark Twain said, 'Never let your schooling get in the way of your education.' I took that to heart."

Appalachian State, paradoxically, has one of the finest environmental studies programs in the country, and Mathis became deeply involved in the school's first clean energy initiatives. Student fees went into building on-campus solar panels, at a time when renewable power was still considered a fringe technology. "It was cutting edge in the country," Mathis told me. "I helped spearhead the first project, transitioning the entire local public transportation system to biodiesel. I worked with them for three years, and had a blast, installing solar and whatnot."

Mathis collects progressive causes like stray cats. He also became involved with the radical feminist movement on campus and entered a graduate program in history and women's studies. During this time he was invited up to West Virginia by a friend who worked with Appalachian Voices, and who was developing the I Love Mountains campaign. Mathis was recruited to assist in a class action lawsuit against coal baron Don Blankenship and his company, Massey Energy.

"What had happened," Mathis explained, "they started dumping slurry into underground mines, pulling out the particulates and putting them in there. The majority of these were made up of heavy metals—it was just a toxic soup that they dumped into the drinking water in Mingo County.

"They had warned Blankenship it was happening and he didn't give a shit. At the same time he built a pipeline to his house, knowing all his neighbors were drinking poison, to ensure that he had clean water. That's who he is."

This was in 2006, just as coal was reaching the precipice off which it has since plunged. Williamson, southern West Virginia, and the plight of the coalfield inhabitants worked like drug substitutes on Mathis's impressionable mind.

"To me, it was a heart-wrenching ethical crime," he said. "It was a real-world experience of all the things I'd been studying and preparing for and thinking about. I was working up on the roof one day with a friend, and seeing Blankenship's house up on the ridge, that's when the vision hit me. The weight of what was going on here, I couldn't turn my back on that."

Mathis saw a chance to test his knowledge in the field of deeds, to build practical applications of political theory. Changing course once again, Mathis shifted his graduate program to Appalachian studies, working under John Alexander Williams, the dean of the field and author of *West Virginia: A History for Beginners*. In 2008, when Obama was swept into office by a tide of youthful idealism, Mathis moved to Williamson and gave himself full time to what is now Sustainable Williamson, a nonprofit combination of environmental activism, economic development, and community empowerment. In the last few years Mathis has worked with the Mingo County Redevelopment Authority and other local organizations to start a Saturday farmers market in a downtown parking lot; he's also worked to install solar panels on the roof of a downtown building that houses the office of Dr. Donovan Beckett, who has joined with Sustainable Williamson to create a health and wellness program to combat the obesity and disease that prey on the people of the mountains. Mathis showed me the modest community garden that he's started, just by the railway tracks on the southern edge of downtown. He's helped push forward the hemp-farming movement, which could provide a sustainable high-value crop that grows readily in the thin soil and on the steep hillsides of the region. And he talks endlessly, to whoever will listen, of his grander vision of parks that combine solar farms, natural gas plants, and industrial facilities on reclaimed mine lands, bringing jobs and prosperity to an area that's been hardest hit by coal's contraction.

Unlike Maria Gunnoe, Mathis has chosen the path of nonconfrontation. He's willing to work with the coal industry to change the dominant paradigm from within. At times he reminded me of the Western aid workers I've met in poor and remote parts of Southeast Asia: earnest and well-intentioned and working with limited resources to right injustices formed over many years of neglect, exploitation, and alienation. The operative myth is Sisyphus. Setbacks are frequent, and small victories prized. Concrete steps forward matter more than newspaper stories, he told me.

"Taking it to streets, chaining yourself to trees, doesn't do jack shit to change the material reality of people's lives," Mathis told me. "We are really chipping away at bona fide economic diversification in the coalfields, which has arguably never been done."

THE MANSION THAT FORMER Massey Energy CEO Don Blankenship used to entertain his corporate friends sits on a ridge overlooking downtown Williamson. You can see its turrets and porticoes from the community garden. Blankenship himself was forced out of the company eight months after the April 15, 2010, explosion at the Upper Big Branch mine, in Raleigh County, West Virginia, which took the lives of 29 men. He has decamped, and the corporate mansion is said to be empty and decaying, a monument to the industry's former glory. Once considered the most powerful man in West Virginia, Blankenship has lived for years under a cloud of possible criminal charges stemming from the Upper Big Branch disaster. The cloud broke in November 2014.

With his Snidely Whiplash mustache and his heavy jowls, Don Blankenship has always looked the part of the evil coal tycoon. Now he may get the chance to play that role in federal prison. The former Massey Energy CEO was indicted in the fall of 2014 on four counts related to the Upper Big Branch mine explosion.

Through two government and two independent investigations Blankenship had scoffed at the notion that he might be held criminally responsible for the disaster, even though two of his former subordinates pled guilty and were sentenced to prison. The indictment charged Blankenship with conspiring "to commit and cause routine, willful violations of mandatory federal mine safety and health standards . . . in order to produce more coal, avoid the costs of following safety laws, and make more money." He was also charged with lying to the SEC and to Massey investors in the months after the accident.

Blankenship's trial promises to provide more evidence of the unsafe practices of the coal industry in Appalachia, and, perhaps, to bring a measure of consolation to the families of the men who died at Upper Big Branch. More than that, it represents the first time the chief executive of a major coal company has been brought to justice for safety violations at his mine. Less than a century ago, in the so-called Coal Wars, federal troops were brought in to quell unrest among miners protesting, often violently, against unsafe and unhealthy conditions in the coal mines. Now a coal CEO faces more than 30 years in prison for allegedly perpetuating, and attempting to conceal, those conditions.

The indictment contained many instances of Blankenship's blunt talk. Officials at mine entrances had code words to warn miners underground of the arrival of mine inspectors and routinely covered up safety violations, per orders from Blankenship. Blankenship berated his managers, in writing, pushing them to avoid spending on safety measures and keep up production.

"Please be reminded that your core job is to make money. To do this, you have to run coal at a low cost, ship your orders, and control your quality." In another message he told managers to ignore proper ventilation requirements for underground shafts where miners were working: "We'll worry about ventilation or other issues at another time. Now is not the time."[13]

"You have a kid to feed," Blankenship told one employee, referred to in the indictment as "the Known UBB [Upper Big Branch] Executive," in a blistering handwritten memo on March 10, 2009—a year before the disaster. "Do your job." Three days later came another handwritten note: "Pitiful. You need to get focused . . . I could Krushchev [*sic*] you. Do you understand?"[14]

After the indictment was handed down on November 13, he was characteristically defiant: "Don Blankenship has been a tireless advocate for mine safety," the lawyer, William W. Taylor III, said in a statement. "His outspoken criticism of powerful bureaucrats has earned this indictment. He will not yield to their effort to silence him."

This time, however, it's not powerful bureaucrats who have accused Blankenship: it's a federal grand jury of ordinary citizens in Charleston, West Virginia, in the heart of coal country. In the justice system, said Senator Jay Rockefeller of West Virginia, Blankenship "will be treated far fairer and with more dignity than he ever treated the miners he employed. And frankly, it's far more than he deserves."[15]

Blankenship is an outlier within the coal industry, which has brought modern technology and new safety regulations to the mines over the last ten years. But for an industry in decline, the sight of a former powerful executive being hauled into court over violations at his company's mines is chilling.

Months before the indictment, Blankenship cordially turned down my request for an interview for this book. Somewhat surprisingly, his counterpart and contemporary Robert Murray, the founder and CEO of Murray Energy, did not, and I spent an hour and a half talking to him, or rather listening to him, by phone in May 2014—a few weeks before President Obama's historic speech on climate change, when he pledged to use the powers of the office to cut carbon dioxide emissions from power plants.

For a man who has spent his life in the coal industry, and who just recently bet $3.5 billion of his company's money on its future, Murray sounds awfully pessimistic. "Contrary to some of the recent coal company statements, there's absolutely nothing on the horizon that makes me think positively about coal demand and prices through 2015 and perhaps beyond," Murray told the crowd at the annual Platts Coal Marketing Days conference in Pittsburgh, in September 2014. "If anything, there's going to be a continued decline in the next few years."[16]

"We are witnessing the absolute destruction of the U.S. coal industry," he told me. "It's not coming back. If you think it's coming back . . . you're smoking dope." The social consequences of the industry's downfall will be simple and stark, Murray added: "Grandma is going to get cold in the dark."

When Hollywood makes a biopic of a twentieth-century coal baron, they will cast someone an awful lot like Bob Murray. He is corpulent, combative, eloquent in defense of indefensible positions, self-made, and very, very rich. Three days after Barack Obama's 2012 reelection, Murray laid off 156 workers, as promised, citing the Obama administration's "War on Coal." Prior to announcing the job cuts, Murray offered a public prayer:

Dear Lord:
The American people have made their choice. They have decided that America must change its course, away from the principals of our Founders. And, away from the idea of individual freedom and individual responsibility. Away from capitalism, economic responsibility, and personal acceptance.

We are a Country in favor of redistribution, national weakness and reduced standard of living and lower and lower levels of personal freedom.

My regret, Lord, is that our young people, including those in my own family, never will know what America was like or might have been. They will pay the price in their reduced standard of living and, most especially, reduced freedom.[17]

He went on to quote 2 Peter 1:4–9: "To faith we are to add goodness; to goodness, knowledge; to knowledge, self control; to self control, perseverance; to perseverance, godliness; to godliness, kindness; to brotherly kindness, love."

How brotherly kindness and love translated into firing 156 of his workers because the incumbent won reelection was not entirely clear, but Murray has made a career of yoking unlikely conclusions to seemingly contradictory premises. He is a contrarian by nature and by business practice. Among the multiple lawsuits he has filed against the federal government for regulating safety and pollution at coal mines is one filed under the Clean Air Act. Even as federal regulations tightened on sulfur pollution from coal smokestacks, Murray dramatically increased his investment in high-sulfur coal mines, paying $3.5 billion for the West Virginia mines of Consol Energy in 2013. In a defamation suit against the *Huffington Post* and liberal blogger Michael Stark, Murray's lawyers claimed that the outspoken Murray—more than anyone else in America the public face of the anti-Obama, pro-coal resistance—was not a public figure at all: "Murray . . . has neither voluntarily sought public or media attention, nor has he achieved such a status by reason of the notoriety of his achievements," reads the complaint.[18]

For such a private person, Murray has a habit of making outrageous pronouncements and grandiose gestures that gain him plenty of notoriety in the mainstream media. He has compared the Obama administration's tactics to those of the Nazi SS. He has publicly opposed the use of personal tracking devices for underground miners, even though he runs the nation's largest underground mining company, one that between 1997 and 2014 paid $18.1 million in fines to the federal Mine Safety and Health Administration (MSHA) for safety violations—including nearly a million dollars after the Crandall Canyon, Utah, mine disaster in 2007, in which nine men were killed in a series of underground collapses.[19]

Murray's crowning moment may have come in April 2014, when the company said it would terminate the medical benefits of nearly 1,200 retired miners who worked for Consol Energy, which Murray had bought the year before. As part of that deal, Murray Energy had agreed to cover the benefits for at least one year. They expired almost exactly one year and a day after the sale closed.

The fault, naturally, lay not with the company but with the president.

In the statement announcing the cuts, Murray spokesman Gary Broadbent claimed that "Murray Energy's inability to provide these benefits is, in part, due to the destruction of the coal industry, including our markets, by the Obama administration and its appointees and supporters, who have eliminated the livelihoods of thousands of coal miners."[20]

Whatever your politics, and whatever you think of large coal companies reneging on their obligations to their retirees, you have to admire the temerity of that statement. And there is no question that the coal industry, even more so than, say, the oil business, breeds such men: outsized in every sense of the word, unafraid of public opprobrium, and convinced of the righteousness of their work and their cause. They could be characters in an Ayn Rand novel. In fact, a reviewer using the name of "Don Blankenship" posted an admiring 700-word review on Amazon's website to *Ayn Rand: A Sense of Life,* a film biography of the objectivist author. Like Rand's hero John Galt, Murray and Blankenship look into the future and see only catastrophe.

"Obama and the Democrats that support him are leading this country into a crisis," Murray told me in his rumbling baritone. "Low-cost electricity is a staple of life—I know because I grew up very poor. The people on the low end of our economic regimen are going to be paying a much higher percentage of their income for heat and for light. As a result, the very people that this president espouses that he supports, he's really making their lives far more difficult.

"One of the cruelest, most evil things I've ever seen a human being do, is when he ordered the World Bank not to fund coal projects around the world. I have great compassion for people who cannot help themselves."

In the real world, of course, the president can't order the World Bank to do anything, even if the United States, as its largest funder, does hold unique influence over the bank's policies. Obama had announced in June 2013 that the United States would no longer finance coal-fired power plants in other countries; the decision by the World Bank, which, like other multilateral funding institutions, has come under increasing pressure to cease backing fossil fuel projects, came a month later.

"Before Obama, coal supplied 52 percent of the energy in America, it's now down to 37 percent. Most experts say it will go down to 35 to 32 percent. I believe it'll go below 30 percent by 2030 to 2035. It can't go any lower because people are gonna freeze in the dark. The Obama administration sees this. But they don't care, because they have to pay back constituencies that got them elected: radical environmentalists, unionists, Hollywood characters, and liberal elitists like Mayor Bloomberg."

I tried to ask about the retirees whom Murray Energy had just cut off, but Murray was just getting wound up.

"There are 90,000 people in the most depressed areas of America that are depending on me. I take that responsibility very seriously. I care about my people deeply. It's a human issue."

That human issue has not kept Murray from battling the MSHA over new rules to tighten the limits on coal dust in mines. The new rules lower the limit from 2 milligrams per cubic meter of air to 1.5 milligrams, still higher than the 1-milligram threshold recommended in 1994 by the National Institute for Occupational Safety and Health.

"I told you that they're not only trying to stop the utilization of coal, but the mining of it too. I hired the best health experts in the world, we did

$2 million worth of studies. We proved that particle size has no influence whatsoever on COPD [chronic obstructive pulmonary disease] or black lung. The researchers also showed that the data the government is using is grossly flawed. It's self-serving, old, inaccurate, and incorrect by region."

I wondered how many months of medical benefits that $2 million could have bought the retired coal miners. Murray again ignored me.

"I said, 'Look, they make these space helmets, called Airstream helmets, with a fan on the back. I'll put every miner in one.' They said, 'No, you don't understand, we want the environment to be 1.5 milligrams,' and I said, 'I do understand: you want to close coal mines.'"

Murray's bluster against "King Obama" has translated into extensive financial support for Republican candidates: according to OpenSecrets.org, which tracks political finance, Murray Energy employees have given nearly $5 million to political campaigns since 1992, virtually all of it going to Republicans.[21] According to a lawsuit filed by a former forewoman at a Murray mine in Marion County, West Virginia, these contributions were not voluntary: Jean Cochenour filed a lawsuit in September 2014, alleging that she was fired after refusing to donate. The filing reproduced a letter to employees, signed by Murray, in which he requested a $200 contribution for each of four Republican candidates for Senate.[22] Murray denies that the contributions were mandatory, but Murray miners were required to attend an August 2012 campaign stop for Mitt Romney at the Century mine in Ohio. Afterward, several of the workers came forward to say that not only were they told they had to be at the rally, but the company shut down the mine for the day and docked them a day's wages.

As if quoting an Orwell tale, Murray chief operating officer Robert Moore told a local radio host, "Attendance was mandatory but no one was forced to attend the event."[23]

As for Murray, at 74 his appetite for public conflict is undimmed. Like Blankenship, he is an incorrigible competitionist. "I work very hard at trying to get the message out as much as I can. I was the CEO of a public coal company once, and public coal company and utility CEOs don't speak out as strongly as they must because they cannot. They have all this diversity in their shareholders, and they have the subject of sustainability, of diversity, to deal with.

"I'm now the head of a private coal company. We never went public, we'll sell close to $4 billion in coal this year, and I can say what I want.

"I do all I can to elect the friends of coal. To push back against those who are opposed to reliable and low-cost electric power. Now what else?"

I SPENT THREE DAYS in and around Williamson, watching the struggle to create a sustainable and prosperous post-coal society. It was not always a heartening experience. At Eric Mathis's suggestion I stayed at the historic Mountaineer Hotel in downtown Williamson. It was one of those places where "historic" is a euphemism for "run-down and outdated." The empty, echoing lobby, with its high ceiling and its faded landscape paintings, had a haunted feel, and the tiny elevator took its time as it wheezed and rattled its way to the fourth floor. The few other guests I saw were retirees on permanent vacation. Carrying my bag down the empty, dimly lit corridor, I couldn't help thinking I'd stumbled into a Kubrick film.

Next door to the Mountaineer is the famous Coal House, a black cube of a building whose exterior is made entirely of coal. Built in 1933 by the railway tycoon O. W. Evans, using 65 tons of bituminous coal from the Winifrede Seam that runs through the surrounding hills, the Coal House was supposed to publicize the "Billion Dollar Coalfield" of southern West Virginia. Today the one-story building houses the Williamson Chamber of Commerce. It was closed the day I tried to visit. The black walls gleamed darkly. It seemed to pull light from the day, a cube-shaped black hole on the quiet street.

On my first evening there we had dinner with Mathis and some of his allies in the post-coal struggle: Jenny Hudson, the director of the Mingo County Diabetes Coalition; Dr. Donovan Beckett, the physician who runs the health and wellness program in Williamson and on whose office roof the solar panels have been erected; and Roger Ford, a local businessman, political consultant, and founder of Patriot BioEnergy, a start-up that's attempting to build biopower plants that burn a combination of coal and biomass from hemp. We met in Pikeville, across the state line in Kentucky, because Williamson has no suitable restaurants open past dark.

"I've been a coal man my whole life," said Roger Ford. "But you can't ignore the market forces. Central Appalachia is just at a competitive disadvantage on a cost per ton basis, versus western U.S. coal and overseas coal. It just is. It's time to face facts."

"People here have spent a long time ignoring the reality of the situation," said Dr. Beckett. "Politicians and the industry, they take advantage of that."

"It's all fear," said Mathis. "It's the oldest play in the book: manipulate people by playing on their fears."

"Well, I don't know," said Ford, leaning back in his chair. A large man with a thick drawl, he'd spent his life in Central Appalachia, but he gave the impression of someone who'd be equally at home in the shark tanks of Washington. He struck me as someone who was used to having his shrewdness underestimated and had learned to take advantage of it. "The industry is not stupid. There's some in the industry, whether they want to be quoted or not, they understand that adaptation to circumstance has to happen. The political world has to take advantage of that and speak for the people who are prepared to face that reality. I'm hoping we can have a more substantive discussion about the future of the region, about the future of coal in general."

That discussion, for now, was limited to the five of us sitting around a table in the otherwise empty restaurant. The gathering reminded me of a clandestine meeting in an occupied country. The headlights of my rental car stabbed into the darkness as I drove back up to Williamson on 119. No one was manning the front desk when I got back to the Mountaineer.

The next morning Mathis and I visited the farmers market, which he and Jenny had helped organize. It was hot in the September sun, and the few produce sellers huddled under a long tent. I counted a dozen tables: peppers, ripe tomatoes, lettuce, peaches, grapes. We sat in the shade and chatted with the farmers and their customers. Traffic was light but the mood was as sunny as the day.

"C'mon in here, big man!" said Doug Dudley, who was selling grapes and peaches, to a friend. At 67, Dudley was hale if slow-moving, and he had a lot to say about the predicament of the coal industry's prospects, little of it good.

"It doesn't look good for 'em this year," he told me between customers. "Your administration is gonna kill coal. The future of the coal industry is not gonna be here. They're gonna eliminate it, unless they come out with some way to burn it efficiently and cleanly."

Dudley was a coal miner for 33 years, in addition to farming 60 acres of bottomland. He went to work for A. T. Massey Coal, one of West Virginia's oldest coal concerns, at age 13. A. T. Massey later became Massey Energy, Don

Blankenship's company, and the owner of the Upper Big Branch Mine. Dudley was a third-shift man: he worked from 11 p.m. to 7 a.m. "I'd catch two or three hours of sleep, get up around 9:30 or quarter till 10, catch the kids one time and then head out to the fields. I'd sleep some later in the day."

He made $13 an hour when he started and $22.50 when he retired at 45. He had little good to say about the younger, improvident miners of today, who can make up to $35 an hour but whose livelihoods are disappearing more quickly than the coal seams themselves. "These younger boys, they go out and buy jet skis and campers, they get themselves in such a jam. Ever' one of 'em's got a $60,000 truck. Why would they do that? It's just amazin' to me."

Dudley's own children escaped the coalfields: his daughter is a physical therapist in Tucson, his son a carpenter in Las Vegas. They don't make it back east much. He spent everything he saved from his mining days to put his daughter through PT school. He lives off his Massey pension, Social Security, plus whatever he makes selling produce. His life is intertwined with West Virginia's past, but Doug Dudley is also part of the future: he's starting to grow hemp.

"Yeah, I got me a few acres," he said, with the caution of a lifelong farmer. "We'll see."

The nonpsychoactive variety of *Cannabis sativa,* hemp has been grown in Appalachia for more than 400 years, since being introduced by the English in the sixteenth century. By the mid-eighteenth century it was one of the biggest cash crops in the region, a $3 million industry with more than 8,000 plantations (most worked by slaves) supplying the cotton industry with twine, bags, and cordage for transporting cotton. The hemp industry declined along with the fall of the cotton industry in the postbellum South, and the passage of the Marijuana Tax Act, in 1937, effectively killed it off, although World War II brought a brief revival of hemp grown for military applications. Today it's seen by many as the center of an agricultural renaissance in Kentucky and West Virginia.

"The potential's there," Roger Ford told me. "Industrial hemp is applicable to a variety of different climates and terrains, it's advantageous for post-mining reclamation, and it can be used as a feedstock that can be easily blended with coal for co-generating electricity."

In 2012 Ford founded Patriot BioEnergy, which is devoted to creating a hemp industry in West Virginia and Kentucky—2 of 12 states that have so far ratified the farming of industrial hemp. Hemp is hardy and drought resistant, and it grows readily in the thin soils of the Appalachian mountainsides—in fact, you can see it today growing wild along roadbeds, called "ditchweed." According to hemp advocates, there are thousands of applications for hemp, from environmentally friendly fabric to rope, paper, oils, cattle feed, and, as Ford mentioned, compressed into pellets for fuel for power plants, blended with coal—thus "co-generation."

Hemp seed oil now sells for about $70 a gallon, and U.S. imports of hemp surpassed $500 million in 2012, according to the Hemp Industries Association.[24] "Our vision is that existing power plants will serve as hubs for integration of agriculture, energy conversion, and manufacturing in a new economy that benefits from the ability to convert biomass, particularly hemp, into thousands of valuable products," said a 2014 report commissioned by Patriot BioEnergy.[25]

"We have the land, we have the water, we have a methodology to get it out with the rail, we have a workforce that could do that," Clif Moore, a Democratic state representative in West Virginia, told TV station WOAY.[26]

Hundreds of plantations employing thousands of former coal miners on abandoned mine lands, growing an environmentally benign crop with a growing market, including fuel for reducing emissions from coal-fired power plants: it's an enticing scenario for economic recovery. There are barriers to this vision, though.

The first is federal law. Despite the growing momentum of the marijuana legalization movement, hemp farming is still technically illegal under U.S. law. A federal permit is required to cultivate hemp, and no permits have been granted since 1999. The Industrial Hemp Farming Bill, introduced in 2013 by Kentucky Republican Tom Massie, went nowhere. In May 2014, federal officials held up Kentucky's first shipment of hemp seeds, imported from Italy, for more than a week at Louisville's international airport. Kentucky's department of agriculture sued the federal government, charging that the feds were trying to hold up the seeds until planting season was over. A last-minute reprieve allowed the delivery; the seeds are now growing in a dozen or so pilot projects around the state. It's not much, but it's a start.

The second problem is that the market for hemp as a source of biomass for energy generation is largely speculative at this point. Biofuels in general, including corn ethanol, have struggled to find viable business models—simply put, using hemp to produce electricity, even in combination with coal, is not economically feasible today.

"I wouldn't come out and say this is a great idea," said Katherine Andrews, the scientific consultant to Patriot BioEnergy. "There are issues of scale and economics. It's really important to find a way to rapidly ramp up hemp crops, to grow them at industrial scale. But if you could, you could reduce the emissions from coal plants; in places where they still want to burn coal it could help marginally."

"Marginal" is not what the world's climate needs. And it's not what the Appalachian economy needs, either. But the lure of hemp—the idea of it as a magic crop that could help solve the region's ills, just as coal was once a magic resource that could erase poverty and ignorance—is powerful, and the first barrier, the obstinacy of the U.S. government, has started to fade. The 2014 Farm Bill, signed by President Obama, included a measure to make pilot hemp programs legal. The bill caused an odd split: on the one hand, law enforcement groups, like the Kentucky State Police, opposed the bill, saying it would make marijuana laws tougher to enforce. On the other side were Republican legislators from West Virginia and Kentucky, including Massie and coal's staunchest ally, Mitch McConnell.

Calling the inclusion of the hemp provisions (which only allow cultivation by universities and state ag departments) "an important victory for Kentucky's farmers," McConnell added that the fledgling hemp industry "could help boost our state's economy."[27] What he didn't say was what many hemp advocates believe: that hemp is part of the post-coal economy in the depressed areas of Appalachia. Other than the brief statement quoted above, McConnell was careful to keep his influence in passing the bill quiet. The move was not lost on the locals, though.

"It's good that Senator McConnell has taken this into the mainstream," Roger Ford told me. "He's trying to preserve his political base as best he can, while also recognizing that we can move forward, that we have to move forward. He thinks strategically, and this is a smart strategic move on his part."

For hemp to move beyond the experimental phase, it needs flat, arable land (hemp may grow on mountainsides, but no industrial-scale agricultural operation can happen there). And flatland among the mountains and hollers and creek beds of Eastern Appalachia is scarce, except for one place: abandoned coal mines.

ONE DAY AT THE CHARLESTON AIRPORT I boarded a small plane to view the coalfields from the air. The flight was arranged by a group called Southwings, which takes journalists, politicians, and business officials on flyovers of coal country to view the effects of mining—in particular mountaintop removal mining, which scrapes the tops of mountains to create flat areas where the coal seams are readily accessible. At the peak, mining companies were setting off 2,500 tons of explosives daily—roughly the destructive force of the atomic bomb that leveled Hiroshima, on a weekly basis.[28]

Mountaintop removal has largely replaced underground mining in Appalachia in the last 20 years, and it has destroyed more than 500 mountains, covering 1.2 million acres.[29] As we lifted off from the airport and headed southeast, 5,000 feet or so up, what we could see was an immense, intricately furrowed and folded carpet of dense, green forest, braided by streams and broken only by a few towns—and by the sudden ugly scars of mines. Our two-hour flight followed a broad loop following U.S. 119, which I'd driven up the day before, then west and north toward the Ohio River, crossing Interstate 64 to circle back east to Charleston. The emptiness of the land was striking; the occasional towns were swallowed by the immense, encroaching forest, part of the most diverse ecosystem outside of the Amazon basin. The mountaintop removal sites were far more visible and expansive than any signs of habitation. They sprawled across hundreds of acres in shades of brown and tan, marked by glistening black ponds of slurry. Much of the removed material would end up in the streams below, in layers of rubble known as valley fill. Valley fills from mountaintop removal sites have obliterated nearly 2,000 miles of Appalachian headwater streams, according to the EPA.[30] It's a brutally efficient way of extracting coal, and it has eliminated thousands of mining jobs. "History shows that the transition from deep to surface mining devastated the region economically, and that the prosperity of mining companies has not gone hand in hand with the economic welfare of coal mine workers," said a 2009 report from Synapse Energy Economics.[31]

The fight against mountaintop removal has become the primary environmental front in the coal wars in the east, and there have been some successes: the Obama administration in 2009 called on federal agencies, including the EPA, the Army Corps of Engineers, and the Department of the Interior to more closely monitor and limit the practice. Opponents have also gained legal victories: in 2012 Patriot Coal (a major coal producer unaffiliated with Patriot BioEnergy, the hemp company) settled a lawsuit filed by the Ohio Valley Environmental Coalition, the Sierra Club, and other groups over selenium pollution from its surface mines in West Virginia. Patriot Coal agreed to pay $7.5 million and to abide by wide-ranging restrictions on mountaintop removal going forward. In the courts, at least, the tide has shifted.

On the ground, the picture is less clear. The Harless Industrial Park is built on a mountaintop removal site; Coalmac is still mining coal at the site. Many people working to revitalize the region assert that the presence of large, open flatlands available for development is a major boon to economic development.

"It's allowed us to do a lot of things we couldn't do otherwise," Terry Sammons, a Williamson lawyer, told me. Those things include the new Mingo County Airport, opened in 2013 on a removed mountaintop outside Williamson, off Mystery Mountain Road. Donated by coal company Alpha Resources, the former mine site represents one of the crowning achievements to date of the Mingo County Redevelopment Authority's plan to bring in new industry and revitalize the region. It doesn't have a lot of traffic yet, but that's not the point, said Steve Kominar, the authority's president.

"Things like this give us the opportunity to go forward," Kominar told me. "When the coal industry started waning, we needed a new strategic plan for Mingo County. We went from the middle to at or near the top in the state in terms of unemployment, and we knew we had to do something other than what we were doing, to start creating opportunities that are unique."

That sort of talk flies in the face of years of abandoned mine-land reclamation, enshrined in the 1977 Surface Mining Control and Reclamation Act (SMCRA, pronounced "smack-rah"), which established the principle of "approximate original contour." That means you restore the land to something like its appearance before you mined it. According to Eric Mathis, approximate original contour results in poor excuses for mountains. And it doesn't come close to restoring the immense fecundity and diversity of the primordial forests. "They want it to look like a golf course," Mathis said as we stood on the mesa of the Harless Industrial Park. One reforestation tactic of the coal companies, encouraged by SMCRA, was to plant autumn olive, which grows in dense thickets on abandoned mine lands across Eastern Appalachia. This hardy Asian transplant, which has silvery green leaves and emits a powerful skunky smell in spring, has invaded pastures from North Carolina to Ohio. Now, land managers are trying to use mountaintop removal sites to reintroduce native trees like black locust and the majestic American chestnut, the "redwood of the East," which was virtually wiped out by chestnut blight in what "many consider . . . to be the greatest ecological disaster of the twentieth century," according to a report from the Department of Interior's Office of Surface Mining in Kentucky.[32]

In 2009, a program called Operation Springboard began planting chestnuts on abandoned mine lands. Since then hundreds of acres have been replanted by teams of volunteers across Appalachia. In the view of foresters and land managers, the effort to re-introduce native species was about much more than restoring biodiversity. Healing the land could help heal the society. "Using the chestnut to promote reforestation efforts," wrote the authors of the Office of Surface Mining report, "on these mined lands . . . brings individuals with opposing views on mining together (miners with environmentalists; teachers with students; elders with children) to engage in conversations about conservation, sustainability, and the future well-being of the region."[33]

IN MINGO COUNTY, those conversations tend to lead back to one man: the late James H. "Buck" Harless, an Appalachian tycoon who had a coal baron's talent for exploiting natural resources and an environmentalists' belief in making them sustainable. In contrast to rapacious indomitables like Don Blankenship and Bob Murray, Harless brought a steward's mentality to the business of mining coal and a neighbor's view of the communities in which he mined it. Born in Taplin, in Logan County, in 1919 to a mother who died shortly thereafter from pneumonia, Harless was raised by his aunt and uncle. After graduating from Gilbert High he did what most of his contemporaries did: hired on at the mines. He eloped at 20 with his wife, June, to whom he would be married until her death in 1999. He worked underground for the Red Jacket Coal company for a couple of years, long enough to save up $500, which he invested in a local sawmill—the first in a long series of canny business deals. After World War II Harless, equipped with a high school education, a keen appreciation for the rough qualities of Appalachian workers, and a born businessman's nose for a good deal, built the company into a regional conglomerate concentrated on coal, timber, and hauling equipment. His regard for salt-of-the-earth West Virginia miners did not include tolerance of collective bargaining or of tree-huggers; the *Wall Street Journal* described him as "the union-battling patriarch of West Virginia's coal industry," and he was a lifelong proponent of mountaintop removal, which he believed, correctly, was the only way for Appalachian mining companies to compete in the world market.[34]

By his fifties, Harless was a billionaire with an estate on an island in the Guyandotte River that included its own helipad. The biggest landowner in Mingo County, he had embarked on his second career as a philanthropist and kingmaker in West Virginia's Republican Party. The fiercely contested presidential campaign of 2000, which pitted Al Gore, already painted by the right as an extreme environmentalist, against the oilman George W. Bush, brought Harless, at 81, his moment on the national political stage.

No Republican incumbent had carried West Virginia since Herbert Hoover, and the Bush campaign was disinclined to put resources into fighting Gore, the scion of a Southern Democratic dynasty, in the state. Harless was convinced that Gore meant to shut down the Appalachian coal industry by banning mountaintop removal (the Clinton administration had supported a lawsuit filed in federal court in Charleston against Arch Coal that effectively halted all mountaintop removal mining in the state), and that coal country would support Bush-Cheney.

The environmentalists' stance against blasting mountains to get at the coal underneath offended Harless's religious sensibilities. "I believe in protecting nature as much as you can," he told the *Journal,* "[but] these things are put here by our maker for our use."[35]

Harless met Bush at a fund-raising event in Austin, and the coal man helped convince the oilman to pursue what became the "coal and cars" strategy: a push to convince voters in the rusting towns of the Ohio River Valley, many of them employed or formerly employed in the auto and coal industries, to vote for a free-market Republican from a right-to-work state.

The effort succeeded; Bush upset Gore in West Virginia and won the state's five electoral college votes—the margin, as it turned out, of his presidential victory. That gave the coal industry a reprieve that lasted well into the first decade of the twenty-first century. Shortly after taking office, Bush reneged on his campaign pledge to take action on limiting carbon emissions. Under Bush, the EPA relaxed its stance on West Virginia's notoriously lax water-quality standards (a decision that helped contribute to the 2014 Elk River disaster, when about 7,500 gallons of a chemical used to wash coal leaked into the river and left nearly 300,000 people without potable water), and the Justice Department declined to pursue further curbs on mountaintop removal. Mining jobs, which had been declining since 1950, rose again. It was morning in Appalachia again, briefly.

But if Buck Harless helped prolong the coal era in West Virginia, he was also enough of a realist to understand that it wouldn't last forever. Today his name is all over West Virginia buildings and institutions, many of which are not monuments to the coal industry: the Larry Joe Harless Community Center in Gilbert, named for Harless's son, who died young in 1995; Harless Stadium at Mingo Central Comprehensive High School (formed by the consolidation of four scattered and faltering high schools across the county); the June Harless Center for Rural Educational Research and Development; and the Buck Harless Student-Athlete Academic Center, these final two at Marshall University in Huntington. Harless was instrumental in the creation of the Hatfield-McCoy Trail System, which threads across the mountainous terrain, much of it privately owned, and draws hikers and mountain bikers from throughout the Southeast. He spearheaded the drive to repurpose mountaintop removal sites as industrial sites, convincing the coal companies that post-mining economic development is in their interest. He was a founder of the Mingo County Redevelopment Authority. And, perhaps most importantly, he fostered the career of

Terry Sammons, who is now among the leaders of the transition to a post-coal future in West Virginia.

Sammons is one of those people whose own biography embodies the tensions and the conflicts that pervade the region. The son of a coal miner, he grew up in Gilbert in the 1960s, when underground mining was just beginning to be replaced by surface mining, with its attendant effects on the land.

"When I was a kid, we lived in Mud Lick Holler; those were the push and shove days," Sammons told me. "Surface mining had come in, and they just pushed everything over into the holler, so when it rained it flooded. I can smell the floodwaters today. It was horrendous but it was just accepted—we were on company property and that's the way things worked."

His father worked when he could find employment in the mines. By the 1970s it was already an unsteady living. "I've seen my dad laid off, go back to work, get laid off again. . . . It was always a frightening time when he'd come home with no work . . . we were afraid at times we were gonna have to leave southern West Virginia.

"This is our home, our heritage and culture. That experience as a child impassioned me on the importance of jobs and sustainability in southern West Virginia. Nothing else is as important."

Resisting the determinism of the coalfields, Sammons and his two brothers managed to get into Berea College in the Cumberland foothills in southern Kentucky. Known as the "poor man's Harvard," Berea charges no tuition to the students it accepts—mostly poor kids from Appalachia. Equipped with a BA in math, Sammons set about trying to figure out how to make a living in southern West Virginia without donning a hard hat and going underground. It was a multipronged endeavor: Sammons taught math, delivered newspapers, sold football tickets, pumped gas. In an encounter that Charles Dickens would have blushed to invent, one day he sold a football ticket to Buck Harless.

"How many jobs do you have?" asked the tycoon.

"As many as it takes," replied Sammons.

"Well, I want you to come work for me."

And that was how Sammons became a coal engineer with successively higher and more impressive credentials: an MS in environmental science from West Virginia University and, eventually, a Harvard Law degree. Sammons, whose firm has offices in Charleston and in Gilbert, has spent most of his legal career representing mining companies, negotiating with environmentalists yet gaining their respect. With his impeccable coal industry bona fides and his long record of bridge-building across the region, he is one of the few people trusted by both sides of the coal versus diversification divide. The longtime chair of the Mingo County Redevelopment Authority, Sammons is spearheading the effort to repurpose mountaintop removal sites as industrial parks and plots for new, hemp-based agriculture.

"Have I run into opposition? Yes," he admits. "There are definitely people who still believe that if it's not coal, they have to be against it. But we should never be at odds. What's happening here is not mysterious. You just have to look around you to see that coal is not the future for this area. And there's no reason, if we do the right preplanning, that these properties can't be positioned for economic development opportunities."

Actually there are reasons, or at least hurdles: one of the main ones is land ownership. If you examine property tax rolls in Mingo County, you will see some familiar names ("Hatfield" and "McCoy"—*those* Hatfields, and *those* McCoys—recur frequently), some lesser-known local families, and a lot of corporations, headquartered elsewhere. For much of the twentieth century these absentee owners were timber, railway, and coal companies. A famous series of articles by Tom Miller, published in the *Huntington Herald-Advertiser* in December 1974, found that absentee owners owned or controlled two-thirds of the privately held land in the state. "Often paying tiny property taxes, they extract the state's rich deposits of coal, timber, oil and gas. And their activities inevitably help sustain the striking paradox of a state with abundant mineral wealth and much abject poverty."[36] Of Mingo County's 271,040 acres, Miller documented, 177,322—65 percent—were owned by four companies: Georgia-Pacific, Island Creek Coal, Cotiga Development, and U.S. Steel—at that time one of the world's largest consumers of coal. It's safe to say that the executives of those corporations did not have the long-term interests of West Virginians at heart.

In recent years those patterns have shifted, but not to the benefit of the inhabitants. Today most big landowners are large holding companies or "resource management investment firms." Two of those firms, Heartwood Forestland Fund of North Carolina and Plum Creek Timberland of Seattle, are the largest and third-largest landowners in West Virginia.

To local activists and anti-coal environmentalists, concentrated land ownership is the unalloyed evil at the dark heart of the region's backwardness, dysfunction, and alienation. But things are rarely so simple in Appalachia. To redevelopment advocates like Sammons and Steve Kominar, the Redevelopment Authority president, having a few big landowners can simplify the process of gaining approvals for improvements like the Hatfield-McCoy Trail System, and, in theory at least, the owners should welcome industrial development on former mine lands that provides an ongoing revenue stream, as opposed to the pure cost of forest reclamation and restoration. Supporting economic redevelopment in West Virginia requires the faith that the Lord's blessings often arrive wrapped in difficulties, like angels in disguise.

"The only way we're going to get things accomplished is to take all the negativity, and say this is where we are, now how can we take advantage of our situation?" said Sammons. "It's a different way of thinking: What do we have that's different than a lot of other areas? What we have is a lot of flat land. Whether you think it's a good thing or not, mountaintop removal has been going on here since 1976 or so, so there's a lot of land already there, waiting to be reused."

"BULLSHIT," SAID MARIA GUNNOE later, when I told her about my conversation with Sammons. The rape of the land was still rape, no matter how beneficial its offspring. "There's an abundance of level land because it was destroyed in order to create what they like to call 'economic development.'"

I thought of this paradox again as I drove north out of the Tug Fork Valley, from Williamson toward Charleston. "If you try to work with the coal companies, the environmentalists say you're in the pocket of the industry," Eric Mathis had told me. "If you try to build a future that's not dependent on coal, the miners say you're against coal." The divide was as deep as an Appalachian hollow and harder to cross. The real legacy of the coal industry in the region went far beyond ruined streams, toxic spills, high unemployment, and startling rates of disability and drug abuse: it reached into the people's hearts and bred fear, mistrust of their neighbors, and a deep alienation from the government that borders on open sedition.

A few weeks after I visited Williamson, the grassroots group Friends of Coal organized a caravan to Washington, D.C., and a rally on the steps of

the Capitol. Reversing the path of the Freedom Riders two generations earlier, a few thousand miners rode buses to D.C. to listen to antigovernment,
anti-Obama speeches, hold up signs that said things like "Proud Miner—
Disgusted American," and generally express their anger at the anti-coal policies
of the current administration. While a few Tea Party activists and a handful
of Republicans came out to witness this show of disaffection and impotence,
it did nothing to slow the market forces that were ravaging the industry in
Appalachia. The onlookers, many of them young professionals who work on
Capitol Hill, were nonplussed by this gathering of hard-hatted and embittered
visitors from somewhere outside the Beltway. The rally was the most visible
expression of the inchoate rage that fuels the miners and that makes outsiders
wonder: Why are they so attached to an industry that has brought them so
much misery?

"It's a disservice to the miners to say that if we just get rid of Obama and
the EPA, then coal's best days are still ahead," Ted Boettner, of the West Virginia Center on Budget and Policy, told me. "It's a disingenuous way of feeding this antigovernment sentiment so that people end up acting, and voting,
against their own self-interest."

"Out here the gravity of discontent pulls in only one direction: to the
right, to the right, further to the right," wrote Thomas Frank in 2004.[37] He
was talking about Kansas, but he might as well have been describing contemporary Appalachia.

That was the question that troubled me as I drove toward Charleston.
How do you convince a people that their nostalgia for a vanishing way of life
is leading to a form of cannibalism, that their kids can't be fed and educated
on rage, that not all change entails betrayal? Raised on misfortune and exploitation, the inhabitants of Appalachia have become masters at playing a lousy
hand. Despair, endured long enough, becomes a form of pride. And no one—
certainly not their elected officials—could convince them that putting down
their drills, their hard hats, and their guns in order to build something new
was going to lead to a better life. "The great war going on in the mountains
comes with too much collateral damage," wrote Nick Mullins, a former coal
miner, on his blog The Thoughtful Miner. "The coal miner and . . . family is
faced with only two options it seems, fight for a side which promises a paycheck

and also destroys their home while causing sickness—or succumb to the poverty . . . running rampant within their mountain home."[38]

The road curved between steep, wooded slopes. The coal tipples and broken-down conveyors became rarer as I headed north. Hollowed by decades of mining, the land arched and curved its back, dark green and watchful. Change was coming, and the people were powerless to stop it.

PART II

THE SURGE

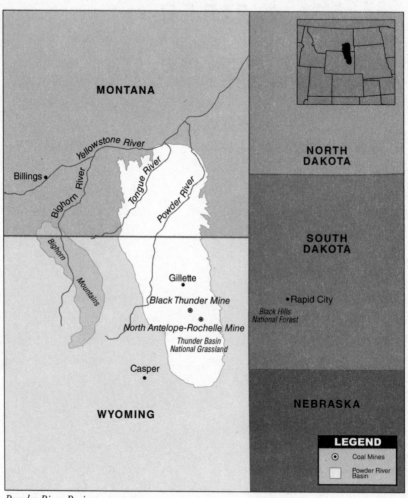

MONTANA

NORTH DAKOTA

SOUTH DAKOTA

NEBRASKA

Yellowstone River

Tongue River

Powder River

Bighorn River

Bighorn Mountains

Billings •

Gillette
•

Black Thunder Mine
◎

North Antelope-Rochelle Mine
◎

*Thunder Basin
National Grassland*

Casper
•

WYOMING

• Rapid City

*Black Hills
National Forest*

LEGEND
◉ Coal Mines
☐ Powder River
Basin

Powder River Basin

CHAPTER 4

WYOMING

I drove up from Douglas, into the Powder River Basin from the south. The weather was changeable, as it often is in the Northern Plains in September; clouds moved slowly across the sky like drifting continents, and the wind bent the tops of the prairie grasses. Along the road were deep oil wells, untended, gas flares bent horizontal by the wind. Disposal sites, rail yards, repair shops, truck graveyards, fuel depots, tire dumps, and, in the distance, the barely visible eastern edge of the Bighorns. The road paralleled the double rail line, and occasionally a coal train, a mile or a mile and a half long, rattled past, heading south, the black pulverized rock humped above the cars in the wind-resistant shape known as the bread-loaf mold. The railway was immaculate, much better maintained than the two-lane highway. Cars are just traffic, but coal is freight and freight is profitable.

The landscape was depopulated. There was industry without laborers or caretakers, as if not only the owners but also the operators were absentee. Here in eastern Wyoming, the mechanization of the energy industry has reached an eerie, empty perfection: we have set this vast machine running, and if this part of the West is any indication, it will go on running when we are gone.

It was odd to see the natural gas being burned off. We are, after all, in the gas age now, a time when America's energy independence is being secured by cheap, abundant gas. Too cheap and too abundant, evidently, to make it profitable here. The owners of these wells are after oil, deep oil, 2,000 feet or more below the surface. Gas is hard to store and expensive to transport. So it burned ceaselessly off into the day, useless carbon vanishing into the atmosphere and adding to the blanket that is slowly, inexorably, warming the earth.

Leaving Colorado that morning, I'd left behind the visible consequences of the changing climate. Boulder, where I lived, had just experienced the worst flooding in a century. The heated summers and less frigid winters had led to drought, and the drought had parched the forests along the Front Range, above Boulder. Millions of square miles of dry standing timber, much of it dead and brown from pine beetles, led inevitably to fire, and three years before, the fire had come to the foothills above Boulder. My family and I were evacuated from our house in Four Mile Canyon for a week. The fire came within a couple of miles of the house but never made it far enough down the canyon to threaten our place. Then, in August 2013, came four days of monsoon rains that, without trees and underbrush to contain them, poured down the canyon sides and deluged the towns of the Front Range. Jamestown and Lyons were almost destroyed; much of north Boulder was underwater, and the basement of our new house in south Boulder, where we'd moved after the fire, had a soaked carpet. We'd gotten off easy. Thus the inevitable chain of environmental catastrophe—from the burning of fossil fuels to the carbon permeating the atmosphere to the bare dry hills unable to soak up the water to the overwhelmed mountain gullies and washes to the rivers of the flatlands bursting their banks—engulfed our little paradise. It seemed like just a foretaste.

As I drove, the scrubby hills rolled to the horizon and an occasional carcass flashed past: usually a fox or coyote, once a pronghorn. In the 110 miles between Douglas and Gillette, the heart of coal country across the Thunder Basin National Grassland, there is the small mining hamlet of Wright, and not much else. With half a million people, Wyoming is one of the least densely populated states. Coming into the Powder River Basin from the south, you have the sense of a great desolation, an immense vacuity. The emptiness, of course, is entirely man-made; less than 150 years ago this land was trodden by vast herds of buffalo comprising millions of animals, not to mention a thriving Plains Indian culture. Covering some 24,000 square miles, the basin resembles, from above, a footprint left by some gigantic mythical creature, aligned north-south and stretching into southeastern Montana. Until the coming of Europeans in the nineteenth century, it was the scene of nearly ceaseless warfare among the native tribes that had occupied the land for centuries. The Crow, the Sioux, and the Cheyenne raided each other's camps and stole each other's ponies until they found a common enemy: the white men. They annihilated Custer and the 7th Cavalry at the Little Bighorn, at the basin's northern end, in 1876, but by the end of that year, after the Battle of Dull Knife, they were essentially defeated and by the following year had been penned on reservations.

Most of the whites who came to the basin before the 1870s were on their way somewhere else. They were headed north and west, to the goldfields of California and the coastal forests of the Northwest, where the fur trade was decimating the population of beavers. Harried by the wind that swept down from Canada, "they saw Wyoming as a high, dry desert to cross and not as a place to settle," writes Dudley Garner and Verla Flores in their history of Wyoming coal mining, *Forgotten Frontier*.[1] The South Pass, at the southwest end of the basin, was a major conduit for trappers, miners, and settlers following the Oregon Trail, which left Fort Laramie, at the confluence of the North Platte and the Laramie rivers, and led west across the Bighorns. From 1840 onward, though, two things began to attract more permanent residents in the Northern Plains—and to fuel the wars against the original inhabitants—buffalo and coal. The buffalo were nearly exterminated within five decades; most of the coal is still underground.

In 1843 and '44 John C. Fremont, not yet the Great Pathfinder, and Kit Carson, already a legendary mountain man, led a series of expeditions that crossed the Bighorns and continued west to the Sierra Nevada range, and they had no trouble spotting the coal beds in "precipitous bluffs" along the North Platte.[2] The coal appeared to be imperfectly mineralized, "and in some of the seams, it was compact and remarkably lusterous [*sic*]," Fremont enthused.

The Indians, of course, had always known of the coal. They called it *cahli* (pronounced "chash-uh-lee") and some of the Plains tribes used it in their sweat lodge ceremonies. Washington Irving, in his somewhat fanciful account of the Bonneville expedition of 1832, claimed that Burning Mountain, on the Powder River, was held in "superstitious awe" by the Crow: "Here the earth is hot and cracked; in many places emitting smoke and sulphurous vapors, as if covering concealed fires."[3] This was among the earliest written descriptions of a burning coal seam. But the natives had no way of excavating it and no ideas for using it if they had.

Guiding a U.S. Army patrol across Wyoming in 1850, Jim Bridger, according to popular legend, found the coal seam near what is today Rock Springs, in southwest Wyoming. The party was commanded by Major Howard Stansbury of the U.S. Army Corps of Topographical Engineers, at that time barely a decade old. Stansbury, a native New Yorker who had never before been west of the Mississippi, proved to be a keen observer. "We found a bed of bituminous coal cropping out of the north bluff of the valley," he noted in his journal, "with every indication of its being quite abundant."[4] The forge at Mormon Ferry, established by the pilgrims of the Latter Day Saints in their trek westward, was fired with coal from the Deer Creek deposit, west of Douglas at the foot of the basin. The U.S. Army began mining coal for its outposts at Fort Bridger and Fort Laramie, and the stations and smithies of the Overland Stage Line were heated with coal. Most of these operations were ad hoc and exclusively local; industrial-scale mining had to await the coming of the railway.

As always in the West, the rail lines progressed along with the exploitation of coal: coal fired the locomotives' steam engines, and the trains brought the miners and carried away the coal. Constructed in the years following the Civil War, the Union Pacific Railroad skirted the southern end of the basin, and the railway owned the mineral rights to thousands of acres on either side of the tracks, not to mention a monopoly over the transportation of the product; the long and tangled relationship between the rail lines and the coal industry started at birth. The Union Pacific "profited handsomely from its coal, using it as fuel, selling it at high prices, and preventing competition by charging outrageous freight rates," according to Wyoming's preeminent historian T. A. Larson.[5] Labor policies were set early on: when miners walked out in 1871, demanding decent working conditions and better wages, the Wyoming Coal & Mining Company fired all the strikers and replaced them with Scandinavian immigrants who worked for $2 a day, well below the pay of the local men. Four years later the drama was repeated at Carbon and Rock Springs: the railroad cut the price paid for coal from 5 cents a bushel to 4, the miners went on strike, and they were quickly replaced by foreign labor, this time Chinese.[6]

Today the railroad no longer owns the coal, the pay is much higher, strikes are few, and the locomotives run on diesel or electricity drawn from distant power plants. The rail line I was following is a spur of the old Union Pacific, jointly owned with Burlington Northern, that heads due north from Shawnee Junction, straight up the spine of the basin, terminating at Peabody Energy's Caballo Mine just south of Gillette. Lines branch into the great open-pit mines to the east: North Antelope-Rochelle and Black Thunder.

The earliest industrial mining operations in Wyoming were at Carbon, west of Laramie, and Rock Springs, in the southwest part of the state. By 1870, two decades before Wyoming became a state, a hundred carloads a week were being shipped east to Omaha and St. Louis. With the pacification of the tribes and the discovery of gold in the Black Hills, settlers began filtering into the Powder River country, and "wagon mines," not served by the railway, began to appear in the basin. By the end of the 1880s, rivals to the Union Pacific had begun extending rail lines into the Wyoming territory: the Chicago and Northwestern reached Douglas, at the basin's southern end, and the Chicago, Burlington, and Quincy Railroad connected Denver with the burgeoning coalfields of northern Wyoming. Along the latter route, in the 1890s, sprang up what would become, in the second half of the twentieth century, the coal town par excellence: Gillette, Wyoming.

"THIS ALL USED TO BE RANCH LAND," Shannon Anderson told me. "Gillette was just a ranching outpost where people came into town to pick up their supplies. None of this was here."

We were driving along Highway 14, north of Gillette. In new suburbs carved from the plains, saplings provided little shade to McMansions that had expensive pickups in the driveways. It was 2013, and the coal industry was supposed to be in serious decline. But it sure didn't look that way in Gillette.

"You see all these new developments, the houses are pretty big—the salaries in the coal industry are good," Anderson said as we rolled back into town. The regional field director for the Powder River Basin Resource Council and a Wyoming native, Anderson had driven over from Sheridan to show me around. Anderson graduated from Grinnell College in Iowa and from Lewis & Clark Law School in Portland. She moved back to Wyoming five years ago. When you come from Wyoming, and you don't work in the energy industry, and you move back to the state as an adult, you feel like you owe the world an explanation. "I have a love-hate relationship with Wyoming," Anderson admitted. "Sheridan's really different than Gillette: it's more diverse, there's more tourism. And you've got the Bighorns."

Gillette, on the other hand, has a new events center with a 9,000-seat arena, an ornate central fire station that puts some Spanish castles to shame, and a public school system with sports fields and gymnasiums to rival a much larger city. Founded in 1892 and originally called Donkey Town, Gillette is a creation of the railroads and the coal mines. While still just an outpost of surveyors' shacks, it was given its more stately present-day name by the grateful bosses of the Burlington and Missouri Railroad in honor of Edward Gillette, the chief surveyor who saved them five miles of track and 30 bridges by rerouting the course of the new line across northeast Wyoming. At the turn of the twentieth century, Gillette housed about 150 souls, no church and no bank. A century later, fed by successive fossil fuel booms, the town had grown to around 20,000 people whose median income, thanks to oil, gas, and coal, was nearly $70,000 a year, well above that of not only Wyoming, but also the United States as a whole. Houses of worship and financial institutions were both plentiful. By the time of my visit the population had increased by half again, to 30,000, and the median income was holding steady. Unemployment was negligible and, according to many in town, a matter of choice. Gillette residents scoff at talk about the death of the coal industry.

So far that reads like the short history of dozens of coal towns across the West. Not many of those, however, have a syndrome named after them.

In a paper presented at the 1974 meeting of the Rocky Mountain American Association for the Advancement of Science, ElDean Kohrs presented a paper entitled "Social Consequences of Boom Growth in Wyoming." "The history of power production—synonymous with 'boom development'—in Wyoming is a dismal record of human ecosystem wastage," Kohrs began. "Frontier expansion without adequate planning has left cities crippled by shameful environments which cause human casualties."[7]

The name Kohrs gave to this dismal record was "Gillette Syndrome," and his portrait of the town was Hobbesian. "Drought, dirt, elemental danger and a dismal battle for existence" characterized the lives of the itinerant coalfield laborers who earned their money in the area mines and spent it in the local saloons. In Kohrs's description, the trailer parks of Gillette were filled with "divorce, tensions on children, emotional damage and alcoholism." Depression was rampant, particularly among housewives, and suicide rates (or, at least, attempted suicides, Gillettians proving mysteriously inept at self-destruction) significantly higher than national averages. The boomtown ethos, Kohrs stated, was "Let it happen; we'll meet the crisis again."[8]

By the standards of twenty-first-century social studies, it must be said, Kohrs's scholarship does not seem overly rigorous. The paper is replete with generalizations about the beleaguered inhabitants of Gillette ("They would not believe that . . . having a 'few snorts' would not cure the ills that beset them"). Kohrs, a psychologist with a PhD from New York University, was a quintessential outsider in Wyoming, and mention of his name, or the syndrome he coined, elicits exasperated snorts and rolled eyes from the locals. And the social ills that he cataloged have certainly dwindled, if not disappeared, in the town's recent prosperity.

Still, the term *Gillette Syndrome* has stuck, and it's hard to deny that Kohrs's appalled response, however moralistic and snobbish, captured something essential about the roughly born boomtowns of the coal era, something that has not altogether vanished even as the churches have multiplied and the saloons have shut down or moved upscale. The coal industry is one of the last in America where someone with no more than a high school education can earn upward of $60,000 a year with full benefits and relative job security.

Gillette and the other energy towns of Wyoming have prospered from a series of successive booms—in recent years roughly one per decade, in Anderson's estimation. In the 1980s, in the wake of the energy crisis of the 1970s, it was oil, as Americans sought feverishly to develop new domestic resources to counter OPEC's might. In the 1990s, after the passage of the 1990 version of the Clean Air Act, it was coal, as the low-sulfur, bituminous coal of the Powder River Basin became highly prized by power generators under newly restrictive federal air pollution rules. In the 2000s came coal-bed methane, a short-lived energy fad that has left abandoned wells across the Wyoming landscape. And now, in the century's second decade, the wheel has turned back to coal and to oil and gas, as new drilling technology opens up vast new petroleum resources across the Intermountain West while coal producers enjoy an unexpected surge in demand.

Sprawled across northeastern Wyoming and southeastern Montana, the Powder River Basin produces about 43 percent of the nation's coal, shipping it to power plants in 200 cities and 35 states. Cities from Tacoma, Washington, to Washington, D.C., and from Duluth to Dallas, run on Powder River Basin coal. Every day nearly 70 trains, carrying a million tons of coal, depart the basin, making the joint line that connects the mines to the main rail corridors to the north and south one of the world's busiest stretches of track.[9] Here coal is so easy to extract that it's cheaper to ship it the 1,400 miles or so to the cities of the East than it is to burn coal from Appalachia. North Antelope-Rochelle, Black Thunder, Eagle Butte, Wyodak, Cordero Rojo: the 13 active mines of the basin produce coal that releases 850 metric tons of carbon dioxide into the atmosphere a year, more than 12 percent of U.S. carbon emissions.[10] In the last several decades—since the passage of the original Clean Air Act, in 1970—the Powder River Basin has become America's energy treasure-house, a seemingly bottomless well that still, in the twenty-first century, provides the primary source for fuel for making electricity in this country.

We stopped the car and stood leaning into the wind. Before us was the Caballo Mine, a relative pup among the vast pit mines of the basin, producing fewer than 10 million tons a year. Behind us, gleaming blue and silver against the dun-colored scrubland, was the Dry Fork Power Station, surely one of if not the last coal-fired station to be built in Wyoming. An occasional big truck lumbered past, buffeting us with its slipstream; the rail line ran through the shallow gully below. In the distance, populating the prairie like gigantic mantises, stood load-out stations, their huge, vertical silos fed by diagonal conveyors. Anderson viewed the scene with a resigned skepticism. The entire lifecycle of the basin's coal, from strip mine to smokestack, was laid out around us.

Near the end of the Cretaceous period, about 80 million years ago, this whole region was underwater. A vast tongue of ocean, known as the Western Interior Seaway or, more poetically, the Niobraran Sea, stretched from the Arctic Ocean to the Gulf of Mexico, covering the middle part of the continent, including the ground on which we stood. Over millennia the water rose and receded, leaving behind dense jungle and coastal swamps, populated by the earliest flowering plants. Giant ferns, conifers, brooding palms, oak-sized horsetails, and 50-foot club mosses, "monsters of the vegetable world," rose above the boggy littoral.[11] As this rank profusion died, fell to earth, and decayed, it formed peat. And under the pressure of millions of tons of overburden the peat was transformed, over millions of years, into coal. The coastal swamps of the midcontinent "would have been like the Florida Everglades, the peat fens of East Anglia, or borders of the Java Sea, which stand just as temporarily," wrote John McPhee in his account of the geology of the West, *Rising From the Plains,* "and after they are flooded by a rising ocean may be buried under sand and mud, and reported to the future as coal."[12] The coal industry of the basin trades on the detritus of vanished forests.

Far to the west, I could see the foothills of the Bighorns humped against the imposing sky. Closer in, to the east, were the Black Hills. Like all of the Rockies, the mountains to the west of the basin were formed in the Laramide orogeny, the great period of mountain building that spanned the late Cretaceous and early Paleocene. In Wyoming, the Wind River Range, the Bighorns, and the Laramie Range all arose, creating, in a few geologic moments, the mountains that define the American West. In the Paleocene they were higher, relative to the rock in which they began their ascent, than any mountain on earth today: 60,000 feet in the case of the Wind Rivers. Almost immediately, those lofty crags started to succumb to water and wind. Erosion and volcanic lava filled the basins between the ranges, until most of Wyoming was a flat surface with a few low hillocks, the tips of gigantic submerged mountains, rising above it. We were standing on the remains of that tabletop now; a second uplift, less than 10 million years ago, had elevated the entire region, excavating the mountains and unleashing rivers that carved deep chasms. This event, known as the Exhumation of the Rockies, created the Northern Plains as they now appear. And it brought the rich coal seams, some of them hundreds of feet thick, close to the surface, where they were easy to excavate.

Indeed, the problem in the Powder River Basin is not finding coal; it's knowing where to most profitably dig. In Wyoming any fool can find coal: 41 percent of the state is underlain by coal deposits. Twelve thousand square miles of the Powder River Basin cover coal-bearing strata, much of it economically mineable. The best coal reserves are often marked by rust-colored outcrops that jut from the grassland like the prows of sunken ships. The rock that forms these distinctive landmarks—called scoria, or clinker—is the remnant of long-lived coal fires that smolder underground. Ignited by range fires, lightning strikes, even spontaneous combustion, they can burn for years, baking and welding the associated shale and sandstone. Softer rock erodes around them, leaving reddish clinker monuments to punctuate the landscape. In the last few million years billions of tons of coal have been consumed in these hidden furnaces.

But there's still plenty left. And the coal majors—Arch Coal, Cloud Peak, and the biggest of them all, Peabody Energy—are scraping it from the earth as fast as they can. The question in the twenty-first century is, Who's going to use it?

WITH A TOUCH AS DEFT as a diamond cutter, Bill Veal controlled a mechanical shovel the size of an office building. Two joysticks—one for each hand—plus a foot pedal maneuvered a toothed bucket, eight feet across, that carved the mine face. The coal was grayish-black, crumbly, ancient. Sitting in the cockpit, three stories up, "One Thing Leads to Another" by the Fixx playing on the stereo, Veal took about a minute to fill the bucket and dump it into the waiting haul truck alongside. Trucks waiting to be filled queued behind the shovel. Each bucket load is 120 tons or so, and each truck can haul 400 tons. Veal's shovel fills a truck bed every few minutes. Twenty-four hours a day, 360 days a year barring mechanical failure or impossible weather. Being in the cockpit of the seven-story shovel—a 4100 XPC electric model from Joy Global, price $10 million minus the electronics that crowd Veal's dashboard—was like riding out a continuous small earthquake. The ground literally shook.

"Ever miss the truck?" I asked over the roar of the machine.

"Nah. Sometimes the sun gets in your eye."

We were in the Elk Pit at the North Antelope-Rochelle Mine, or NARM, the vast mine complex south of Gillette owned by Peabody Energy. Veal's was one of two shovels working this face. He was in the middle of his 12-hour shift and, at 61, nearing the end of a 40-year career in the mines. Running a shovel is at the top of the pyramid in the mines, and Veal is one of the best. In fact, he was working on a new record. Last year, his shovel, No. 111, set the all-time NARM record for material moved: more than 38 million cubic yards.

"That's more than we've ever done," his boss, Scott Durgin, told me later. "I think it's a world record but I don't want to say unless it's in Guinness."

Like most of the miners at NARM, Veal comes from elsewhere: "I'm just a farm kid from South Dakota." White-haired and gnomic, he makes a good living: an A-tech, his level, earns $32 an hour plus overtime. The labor unrest that pocks the history of the coal mining industry is a thing of the past, at least in Wyoming; these jobs are coveted, often passed from generation to generation, and Peabody prides itself on treating its employees well. The end of coal will be unwelcome news to the people of the basin.

"I imagine I'll do it another two or three years," said Veal, taking a brief break from running the shovel. "There's always a lot of challenges, trying to get it done in a certain amount of time, to keep the face clean. I'm still learnin'." Veal's timing is good; he'll be retired before change comes to the industry in the basin. And as he sets earth-moving records, NARM is running flat out.

Every day from NARM's twin load-out facilities, 20 trains are filled with coal and sent to points east, west, and south. At about 125 cars a train and 110 tons of coal per car, that's 275,000 tons of coal a day, from this one mine. Coal production at NARM topped 107 million tons last year and was expected to reach 113 million in 2013—topping the all-time record of 109 million, set in 2011. If the coal industry is expiring, it's putting on a hell of a death scene in eastern Wyoming.

Even as the United States weans itself off of coal, hunger for the world's dirtiest fuel has continued to grow, not only in the developing countries of Asia—particularly India and China, where rapid economic growth has fueled a seemingly unlimited hunger for coal—but also in some First World nations. Coal demand has actually risen in the last two years, as developed countries like Germany and Japan phase out their nuclear power plants, and power generation from renewable sources, like solar and wind power, continues a gradual ramp-up that could take another couple of decades. Coal-fired power generation in Europe rose 18 percent in 2012. Efforts to shift China's power generation away from coal and toward cleaner sources are starting to take effect, but they've only slowed the growth in coal use, not reduced it. China, which only became a net importer of coal in 2009, now consumes nearly 40 percent of the world's coal. Every new power plant built in China is a boon to the world's coal producers, like Peabody.

According to the International Energy Agency, coal consumption worldwide, driven by the expanding power sectors in China and India, will increase by around 1.4 billion metric tons over the next five years, even as U.S. demand declines.[13]

The U.S. Energy Information Administration estimates that world energy demand, barring some dramatic increase in energy conservation, will rise by 56 percent from 2010 to 2040.[14] Coal's share of that energy will drop, but only by 1 percent or so. U.S. producers exported a record amount of coal in 2012. Even here in the United States, coal has rebounded as the price for natural gas clawed its way back from the historic lows of 2011 and 2012: after slipping to 37 percent in 2012, coal's share of U.S. power generation climbed back to the 40 percent range in 2013–2014.[15] Coal mining companies like Peabody, whose shares have lost three-quarters of their value since April 2011, saw a slight recovery in late 2013–early 2014. The coal surge inspired Peabody CEO Greg Boyce to predict a "supercycle" for coal, where rising prices for natural gas, booming demand for coal in the developing world, and the advent of new technologies for "clean coal" will combine to raise the price of coal and restore the black rock to its rightful place as the world's number-one source of electricity.

That's the vision that drives Bill Veal's hands on the joysticks of the XPC shovel, and the lineup of haul trucks winding through the pit to a central crusher and then on to the load-out, where the train cars are filled. Wishing him luck, we climbed back into the white Peabody SUV and drove eastward, through the gray chasms marking the pits among the tan folds of the plain. We stopped at Elk Pit, a younger site, where a gigantic dragline was moving rock and dirt that had been blasted off the pit wall a day or two before, to expose the coal seam.

NARM has four draglines, and they embody all the brute force of strip mining. The free-swinging bucket can scoop 135 cubic yards of material in a single grasp, enough to fill a hundred pickup trucks. On its boom, which is longer than a football field, the bucket swings in a 180-degree arc, like a pendulum, digging on alternate sides to conserve the inertia of the moving boom, saving both time and energy. A cockpit and control center the size of a small house sits at the base of the boom. The whole thing looks like a gigantic mechanized dinosaur, scanning the plain for prey. Clouds of dust rise as the bucket empties, and a companion dozer, tiny by comparison, scurries to pile the overburden in tidy mounds. The draglines, which move when they need to at a mile an hour on retractable shoes, are electric, too: the world's biggest extension cord snakes behind it to a nearby diesel generator. Driven by the coal mines, which use tremendous amounts of electricity to produce coal that will be burned to produce more power, Wyoming uses more electricity per capita than any other state.[16] The price per kilowatt-hour that Peabody pays for that electricity is well below the national average. Coal is cheap to mine, as long as electricity is cheap.

Durgin, who manages NARM along with Peabody's three other mines in the basin, gapes at the operation alongside me. Another South Dakotan, Durgin, at 38, conveys the easy competence of an engineer turned manager. He has witnessed the explosive growth of this mining complex for two decades, but he has lost none of his wonder at the vastness of the mechanism he oversees. "Impressive, isn't it?" he asked me as we watched the dragline clawing the earth.

Durgin's enthusiasm is born, at least partly, out of faith. Many of Peabody's employees in the basin are, like Veal, nearing retirement age; Durbin is in the middle of an ascendant career, and he has three young sons at home. His stake in the long-term future of coal is driven as deep as the coal pits. To a man in Durgin's position, the idea that all this—the open pits, the draglines and shovels, the crushers and load-outs, the trains snaking endlessly away to the cities of the coasts—might melt into air within his working life is not just incredible; it's unthinkable. He is as invested in the mining of coal as it's possible to be. His own future and his family's future are wrapped up in the continued health of the coal industry, and of Peabody Energy in particular.

That health depends not upon plants in Dallas or Detroit but on ones in Dongguan and Dalian. The future of coal-fired power generation—and of metallurgical coal, used to fire the furnaces that smelt iron for steel—lies in Asia.

ALREADY CHINA CONSUMES more than half of the world's coal production. With an economy growing at more than 7 percent a year, China has hundreds of millions of people eager to rise into the middle class—and to acquire the refrigerators, air conditioners, and flat-screen TVs that mark that status. That means the country must continue to increase its energy output dramatically over the next 20 to 30 years to fulfill the dreams of its people and the ambitions of its government. Coal demand in India and the rapidly advancing nations of Southeast Asia, such as Indonesia and Vietnam, will also expand rapidly in the coming years, unless other forms of inexpensive and accessible energy can be developed. But China, increasingly, is the engine that drives the worldwide coal industry.

To be sure, China is embarked on a far-reaching program to decrease its reliance on coal, through a renewable-energy program that is as ambitious (and well funded) as any country's in the world, and through a tough set of regulations aimed at increasing energy efficiency and reducing energy intensity (i.e., the amount of energy consumed per unit of GDP). In Part III of this book I will report on those efforts firsthand. Nevertheless, many experts believe that the rise of coal use in China is inexorable and that there is little the government, or Chinese activists, or outside organizations can do about it.

"It is very unlikely that demand for thermal coal in China will peak before 2030," said William Durbin, the Beijing-based president of global markets for international consultancy Wood Mackenzie, in a statement accompanying the firm's June 2013 report, *China: The Illusion of Peak Coal.* "Why? Because China's aggressive investment program for nuclear, natural gas and renewables capacity is centered in the coastal region, while coal-fired capacity grows in the central and western provinces."[17]

Coal demand in China will double by 2030, Wood Mackenzie forecasts, growing to 7 billion metric tons (7.7 billion short tons, the type used in the United States) a year of thermal coal. That is just below *total world coal demand* for 2012. In other words, global coal output would have to increase by more than one-quarter to meet China's growing demand. To say the least, that is a prospect that chills the hearts of climatologists. For executives at Peabody Energy and other Big Coal producers, on the other hand, it represents a winning long-term business strategy.

I spoke to Greg Boyce, the CEO of Peabody, at his office in downtown St. Louis on a raw, snowy day in January 2014. Outside the windows of his spacious corner office, the top of the Gateway Arch, on the bank of the Mississippi, disappeared into clouds. Boyce greeted me cordially. He looks exactly like what he is: a mining engineer turned executive, with the ruddy cheeks of a man who's spent a lot of time outdoors, at mines around the world. For someone who heads a corporation that is vilified by many environmentalists for perpetuating a fuel that is slowly—or, not so slowly, if you pay attention to the latest reports from the Intergovernmental Panel on Climate Change—choking the planet with carbon dioxide, he doesn't seem to lose much sleep over what he does. In fact, Boyce, who made nearly $11 million in 2013 in salary, stock, and other compensation, seems like a guy who sleeps quite well at night.[18]

For many years Peabody was at the forefront of climate change denial. Fred Palmer, the company's combative chief lobbyist—Dick Cheney to Boyce's George W. Bush—is best known for saying, in an interview in the late 1990s, that burning fossil fuels and pumping carbon dioxide into the atmosphere is "doing God's work." "Any actions that would reduce coal-fueled generation harm Americans," Peabody said in a statement responding to the president's June 2013 climate change speech.[19]

Recently, though, Peabody has largely abandoned the climate change battlefield. In public statements Boyce has pivoted to a less controversial position: coal has provided light, heat, and power to millions of people, improving living standards for nearly two centuries, giving rise to the greatest increase in living standards in human history. Millions of people are still without access to affordable, reliable electricity. We need coal more than ever.

"Our conversations around climate change haven't really changed," Boyce insisted. "What we've done is try to focus the discussion around the fact that we've got a global energy poverty problem. No matter what scenario you look at in terms of the climate, unless we make inroads at solving energy poverty, the environmental issues will never be solved. You need people worrying about something other than how they're going to put food on the table before you can tackle climate change. That's the number one challenge. How do you meet the requirements for energy for the whole world? Coal is the only solution."

The market, I suggested, has decided otherwise: prices continue to fall, along with the share prices of coal companies, Peabody included. What would it take for the industry to recover?

"What's it going to take for prices to rise again? We've seen massive amounts of [mining] projects canceled. At some point, the continuous demand [will] eat up the excess supply, we'll see another lag before new supply can be built, and we'll have another rise in prices. Then you'll see what our platform can deliver in a rising cycle."

The global seaborne market was expected to exceed 1.3 billion tons in 2014 for the first time ever, Boyce told me. "New coal-fueled power plants and steel mills are rising along the coasts of India and China, where populations are expanding most rapidly, and these plants need seaborne coal. China and India are increasing their coal imports at double-digit rates. Some 15 to 20 million more people will move [to growing cities] each year over the next decade. And all of these people will require more and more energy just to meet their basic needs. You can't get from point A to point B with coal on the decline. It's not possible."

"Is it possible to get to point B, using coal, without destroying the resources all those people need to survive?" I asked.

"Look, the climate has been changing since the day the earth was formed. The question is, as we see an increase of carbon dioxide in the atmosphere, what does that translate into in terms of how we address energy and economic activity? Our perspective is that every day we use coal we ought to use it cleaner than we did the day before."

Peabody, Boyce said, is determined to supply this market through multiple channels. "We supply Asia from Australia; we supply Asia in-country through projects in Mongolia and China; we supply Asia through our trading and brokerage group, which specializes in physical delivery of third-party coal; and we are looking at exports through the West Coast to deliver the cleanest coal in the world to Asian ports."

The last leg of that four-legged table—exports to Asia from the U.S. West Coast—has helped galvanize the anti-coal movement in the United States. For many years the United States has shipped scrap and garbage to China—more than $11 billion worth in 2011, according to the U.S. International Trade Commission[20]—and now we want to ship our dirtiest fuel to China, while shutting down coal plants at home. An unlikely but vocal coalition, comprising Native American tribes, ranchers, environmentalists, and local officials, has formed to block the construction of new export terminals on the West Coast. A series of contentious public hearings around the region in 2012 and 2013 provided a highly visible stage for environmentalists and local officials to passionately argue against the economics and the ethics of shipping America's coal to be burned in Asia.

At one point, three years ago, there were six coal export terminals planned or proposed on the West Coast. Three of those had been scrapped as of late

2014; the remaining three face vigorous local opponents, years of environmental impact assessments, and questions about financing.

"Seattle businesses, residents, and property owners would face unacceptable impacts if coal export facilities are built in the Northwest," said then–Seattle mayor Michael McGinn in August 2013. "Beyond Seattle, it's critically important that we continue to educate ourselves and others on the global implications of burning coal—a major contributor to local air pollution and worldwide climate change through the effects of greenhouse gasses."

(McGinn was defeated for reelection in November 2013 by Ed Murray, a state senator whose campaign received strong support from business groups and from the coal industry. A few weeks before the election, however, Murray also came out against the expansion of exports, saying he would continue McGinn's struggle against huge new coal terminals on Puget Sound.)

Opposition to coal exports stems from a curious combination of NIMBY-ism, climate change fears, and concern over the morality of, in effect, shipping dirty power generation to China. Coal industry officials scoff at some of these arguments, such as the notion that increased coal shipping would increase the public health damage from coal dust blown from open rail cars; and it's true that the coal industry has spent millions on surfactants—spray-on substances that minimize escaping dust. What's not in doubt is that increased exports would vastly expand rail traffic from the basin to the coastal cities of Washington and Oregon. Coal traffic from Wyoming to the Northwest would double if the proposed exports go through, according to some studies. According to a study by environmental engineering firm Parametrix, commissioned by the Washington State Department of Transportation, congestion at rail crossings would cost local businesses $384,000 to $455,000 annually, and real estate values along the coal train route would fall by $270 million to $475 million along the tracks.[21] Such knock-on effects are notoriously hard to quantify. The coal industry responds, predictably, with figures citing the economic benefits of new coal export terminals: Nearly 12,000 new jobs, more than $3 billion in additional tax revenue and indirect economic activity.[22]

The underlying reason for opposition, though, is based not on economics or local environmental harm; it's rooted in the conviction that the more coal we export to Asia, the more coal Asia will burn; and the more coal burned in Asia, the less chance we have of stemming runaway global warming.

"People are getting involved from all sorts of perspectives and viewpoints," Ross McFarlane, a senior adviser at Portland-based Climate Solutions, says of the anti-coal movement. "But many communities and individuals are looking at it from a broader perspective and asking, 'Is this a good thing for our state, our economy, the planet?' And they're concluding the answer is No."

This, to put it mildly, is aggravating to coal industry executives.

"The notion that curbing U.S. coal exports will in any way curtail coal use in Asia is wrong in purpose but also wrong in practice," Boyce told me in an email. "China, India and others will import all the coal they need from other eager nations, such as Indonesia, if the coal doesn't come from the United States. And many competing nations lack the environmental controls and safety practices that we have here. So, the question is not whether Asia will import more coal. It will. The question is whether the jobs and economic benefits will accrue to the United States or not. When people understand the question, the answer seems easy."

In 2010, at the twenty-first World Energy Congress in Montreal, Boyce unveiled what he calls the Peabody Plan to burn more coal to bring electricity to the billion or so people who still live without it, create jobs, and develop clean coal technology for the future. "The greatest crisis we confront in the 21st Century is not a future environmental crisis predicted by computer models, but a human crisis today that is fully within our power to solve," Boyce said in his Montreal speech. "Only once we have a growing, vibrant, global economy providing energy access and an improved human condition for billions of the energy impoverished can we accelerate progress on environmental issues such as a reduction in greenhouse gases."[23]

Peabody sounded a similar note in its 2010 response to the Environmental Protection Agency's plan to regulate greenhouse gas emissions (GHG, in Peabody's shorthand) from power plants: "Because GHG emissions, particularly carbon dioxide emissions, are so closely tied with all facets of modern life, a finding that GHG emissions endanger public health and welfare is akin to saying that modern life endangers public health or welfare," Peabody wrote in its petition to EPA. "But plainly just the opposite is the case."[24]

It's an impressive feat of rhetoric: allying the interests of one of the biggest extractive companies in the world with the poor, lightless masses of Asia and Africa. However you feel about coal miners, though, it's inarguable that the developing world needs low-cost power. And the shale gas boom, the rapid fall of solar power prices, and the long-delayed "nuclear renaissance" notwithstanding, coal remains the lowest-cost, and often the only, source of power in most regions, certainly in the booming economies of Asia.

Peabody is hardly tying its future to exports from the United States alone. In 2006, in the pre-crash coal boom, Boyce engineered a $1.5 billion acquisition of Excel Coal Ltd., one of Australia's largest producers of metallurgical coal (commonly called "met coal" and used in steel production and other industrial processes), giving Peabody a vital doorway to the booming China market. He followed that with an even larger strategic takeover, the nearly $5 billion purchase of Brisbane-based Macarthur Coal, one of the last remaining independent mining operations in Australia.

The Australia acquisitions propelled Peabody, which began in the 1880s as a small Chicago operation that bought up supplies from surrounding mines and sold it to homes and businesses in the city, on its way to becoming the world's most powerful and profitable coal supplier. And they pushed the company's share price to a peak of nearly $73 a share in March 2011. Today, Peabody's Australia properties contribute 49 percent of the company's revenues, and one-third of its profits, mostly supplying met coal to China's steelmakers.[25] Peabody has also moved into China's domestic coal market, partnering with the government of Xinjiang, the politically restless region in the country's far west, on a huge coal mine project that could ship 50 million tons of coal a year to the booming cities of the coast. Peabody is also one of the leading Western companies working to develop the vast Tavan Tolgoi mine—site of the richest met coal reserve in the world—in Mongolia.

The Australian mines, the in-country China and Mongolia projects, the coal traded through its in-house brokerage (known as the Coal Trading Group), and, of course, the expanded exports from the U.S. West Coast, all are parts of "a holistic strategy of serving these growing markets," said Vic Svec, Peabody's senior VP of investor relations. "We believe over time that exports from the Pacific Northwest will occur; we're talking about a product that is legal to mine and use in the U.S. and other nations. There's no legitimate reason to exclude its export from any place."

This whole holistic strategy, though, is dependent on one key assumption: that coal use in China will keep rising indefinitely. Not everyone considers that inevitable. In early 2013 the People's Republic launched its first carbon emissions trading program. A few months later, analysts at Goldman Sachs, foreseeing a "sharp deceleration in seaborne [coal] demand" (with annual growth declining from 7 percent in 2007–12 to 1 percent in 2013–17), stated that "the window for profitable investment in coal mining is closing."[26]

In September 2013 Citi Research released a report titled "The Unimaginable: Peak Coal in China," which argued that a combination of factors—including intensifying efforts to reduce China's crisis-level air pollution, the slowing of economic growth and the reduction of energy intensity (the amount of energy used per unit of GDP), accelerating use of clean energy and natural gas, and efficiency improvements in existing power plants—may well lead to "the flattening or peaking of thermal coal demand for power generation in China by 2020."[27]

That could scupper the strategy of not only Peabody, but also of other Big Coal companies, such as Arch Coal and Cloud Peak Energy, along with the Big Coal-export industries of Australia and Indonesia. In October 2013, mining giant BHP Billiton—which, unlike Peabody, is a diversified conglomerate with aluminum, petroleum, and copper mines—canceled plans for a big new coal terminal in Queensland, on Australia's northeast coast, as well as the rail lines to bring the coal to the sea.

In other words, the fight over coal exports represents, at a basic level, a struggle over the future of the coal industry. Bruce Nilles, the senior director of the Sierra Club's Beyond Coal campaign, told me that if the China gambit fails, "that's the end of the story" for Peabody and other U.S. coal producers. "Without that they have no story."

At NARM, Peabody has a 60-year plan for continued mining. The coal left in the ground in the basin is, according to the company, effectively limitless. If exports to Asia are blocked, the market for that coal may be limited.

"Someone described the coal companies as the box turtle population in Virginia," Nilles said, "where there are only adults left—adults who are too old to mate. They're extinct, but they don't know it yet."

EXTINCTION TAKES MANY FORMS, and the human condition demonstrates that an awareness of your impending demise is no guarantee of salvation.

According to the geographer and writer Jared Diamond, our species is unique in our ability to make "counter-evolutionary choices"—such as, for example, continuing to strip the Powder River Basin of its coal reserves at a time when simple arithmetic proves that burning all that coal is sure to produce a desiccated world where large regions now populated become uninhabitable.[28] That category is certain to include the arid high plains of eastern Wyoming—where more than 3.5 million acres of forest have been infested with tree-killing pine beetles (an infestation caused by mild winters that allowed the pine beetles to survive and multiply); where, thanks largely to oil and gas production, ozone levels in some parts of this lightly populated state routinely surpass those in Los Angeles; where water shortages in the North Platte Basin have already become acute; where hundreds of abandoned uranium mines, left over from a previous period of energy development, dot the land, many within sight of the major coal mines; and where less than 6 percent of the land stripped for coal mining has been fully reclaimed.[29] That these are the natural consequences of human choices, that all of this is directly related to the gigantic machines scraping coal out of the earth and the trains carrying it to distant plants, is not lost on those who are feeling the effects today.

Among them are ranchers who have been cast at times into an uneasy alliance with environmentalists in decrying the effects of unlimited, largely unregulated coal mining. Now they're asking what comes next. For L. J. Turner—whose grandfather arrived in the basin shortly after World War I and who today runs several hundred head of cattle and sheep on land that is steadily being encroached upon by the mines—those effects can be seen in the creek that runs just past his house. Or that used to run.

"Where we used to have a lot of fish holes, spring holes, now we got mud holes," Turner told me. "If it's wet enough to make mud."

Not just the surface water is drying up, he said: the underlying groundwater, part of the great High Plains Aquifer system that stretches from Wyoming and South Dakota south to Texas, is dropping as well.

"We've lost half a dozen water wells that were drilled, just in the last ten years. They're expensive to drill a couple of hundred feet, but if we lose this aquifer, we'll have to go down a couple of thousand feet. You're talking about more money than what you can afford to do."

We were talking in the kitchen of Turner's house, which sits in a shallow draw an hour or so south of Gillette. Filtered by cottonwood branches, golden light fell on the polished wood beams of the floor, the walls, and the table at which we talked. At 72, Turner is weathered but hale, with the rough hands and plaid work shirt of a gentleman rancher but the ready indignation of a practiced scrapper. He seemed like a man who had settled most of the big questions in life to his own satisfaction.

"The surface aquifer has been knocked down five feet. All these little creeks through here, they might have had five feet of water in them in the spring. There's nothing there now."

Both of Turner's grandfathers homesteaded in the basin in the early part of the twentieth century, traveling west on the "transcontinental rut"—the unpaved trail cut over a century by wagon trains, and later motor vehicles, across the High Plains. "They come up from Cheyenne through Douglas and Ross. I think there was one bridge they crossed, the rest they had to ford."

The ranch established by his ancestor now covers 10,000 acres of rolling grassland. L. J. and his wife, Karen, built the house in which we are sitting in 2000, just down the hill from the old ranch house and barns, which had burned more than once in the previous 50 years. The Turners raised four children here, now scattered. One son died at 21, in Bosnia on a NATO peacekeeping mission. One daughter, a wildlife biologist who specializes in the study of anthrax in animal populations, is working in Oslo, from where Karen had just returned. A second daughter lives in Connecticut, another son in Colorado. While we talked, Karen brought in an armload of vegetables from the garden outside the house. It was mid-September, and during her three-week visit to her overseas offspring, it was L. J.'s responsibility to begin the harvest.

"I think you missed one," she remarked drily, placing a zucchini the size of an axe handle on the table. "Didn't you pick any of 'em?"

"I did!" L. J. protested. "There's some in the refrigerator!"

In addition to the privately owned grazing lands, Turner for years ran livestock on federal land, on the Thunder Basin National Grassland, which covers most of the southern half of the basin. Many of the mines in the northern basin are privately owned; the ones to the south, including the two giants, Black Thunder and NARM, are on federal land, leased from the Bureau of Land Management (BLM) or the National Forest Service by the mining companies.

Turner and his family have grazed cattle and sheep on government grassland since the 1930s, under favorable lease terms. Then, a few years ago, that changed.

"The Forest Service told us that Peabody was expanding their mine, and we wouldn't have the [grazing] permit anymore. I was in the Thunder Basin office down at Douglas, and the lady there said, 'You know you're not gonna be able to use this permit up there next year.' That's all the notice I got."

Over the last several years Turner has lost access to about 6,000 acres of grazing land, he said. Between the water shortages and the lost pasturage, he's had to sell off most of his cattle herd. Before, he raised both commercial cattle, sold to slaughterhouses for meat, and registered cattle, used for bloodstock. Now he runs only registered. His inquiries with the Forest Service as to the exact nature of the process by which he forfeited long-held grazing rights, while the coal mines continued to expand, went nowhere.

That's not unusual; the entire mining rights leasing program, run by the BLM, has for years been shrouded in a veil of opacity, controversy, inside dealing, and mismanagement. It's a remnant of the days when mining rights were conferred and traded in county courthouses and private dining rooms, with little regulatory oversight and less public scrutiny.

The shortcomings of the coal leasing program are hardly a secret; 2013, in fact, was a banner year for reports, most of them damning, that exposed the questionable nature of the program and the cost to taxpayers. In July 2013 the Office of the Inspector General of the Department of the Interior—in other words, the watchdog within the federal agency that administers the program— "found weaknesses in the current sale process that could put the Government at risk of not receiving the full value for the leases."[30] That's bureaucratic language for a decades-long giveaway in which coal mining companies have paid pennies on the dollar for the true value of the coal they remove from Wyoming and other Western states. This is not new information. The Government Accounting Office found that a series of sales of coal leases in the basin brought in $62 million for the Interior Department, $3.5 million less than the department's original estimate of the coal's value. In fact, the GAO concluded, the fair market value of the leases was $100 million more than what the mining companies paid. That report, which covered only two lease sales, came out in 1983.

In the case of the basin, all of this has occurred on land that the federal government, managing to ignore the obvious presence of one of the world's largest mining complexes, considers unfit for coal mining. In 1990 a little-known government fiefdom known as the Powder River Basin Regional Coal Team (comprising three federal bureaucrats and two local officials) decreed that, for purposes of consideration under Interior's leasing program, the basin, source of 40 percent of U.S. coal production, is "outside coal production regions."[31] This Orwellian ruling effectively eliminates the federal government's role in deciding what tracts of land should be leased and how individual sales fit into the larger regional landscape. Known as "lease by application," this congenial arrangement also eliminates competitive leasing, because individual companies spend years, and millions of dollars, on exploration and geological studies to design and designate the properties they want to mine. "Under the current system, industry decides whether, where, and how much coal they want to lease, subject to BLM approval," writes Mark Squillace, a former professor of law at the University of Wyoming, in Laramie, now teaching at the University of Colorado.[32]

Since 1990 the government has leased nearly 8 billion acres under this regime, much more than most economists believe that market could bear, at prices far below the actual market value of the coal, on terms essentially dictated by the mining industry. The amount lost to U.S. taxpayers through this congenial arrangement has been estimated at $28 billion.[33]

It's useful, here, to think of coal as a depreciating asset. The 12-year coal price chart, from 2001 to 2013, resembles a relief map of Denali: foothills in the early 2000s climbing to a major peak in 2008, and then descending into foothills again thereafter.[34] In June 2008, two years after Greg Boyce took over as CEO of Peabody, the price of coal hit nearly $140 a ton. It's been downhill ever since. The price has still not reached the trough of 2001, when recession, record production levels, and tightened restrictions on sulfur emissions drove coal below $25 a ton. But the trend is clear: coal lost 20 percent of its value in 2013, dropping below $60 a ton, and analysts forecast further slides ahead: coal markets are oversupplied, wrote commodities analysts at Bank of America/Merrill Lynch in late 2013, and "the outlook is everything but rosy."[35]

Federal and state regulators left billions on the table when coal was at its peak. Now that the price is sliding—probably for good—no amount of coal lease reform is ever going to recapture that value. Ultimately, the argument over the leasing program is part of a larger argument over the future of coal in the twenty-first century. And that argument has been going on for a very long time.

THE FIRST DIESEL LOCOMOTIVES went into service in the 1930s, replacing coal-fired steam engines, but it took until after World War II for the full diesel-ization of the railways to take hold. In 1947, two days before Christmas, Union Pacific shut down its last coal mines in Wyoming. The last standard-gauge rail line to use steam engines, the Leadville branch of the Colorado and Southern line, shut down its last steam-powered locomotive in October 1962, and the coal industry spiraled into a depression from which it didn't fully recover until the 1970s, when new air pollution standards made the low-sulfur coal of the Powder River Basin much more attractive. It's a little-mentioned fact that the 1970 Clean Air Act, one of the most successful pieces of environmental leg-islation in American history, also helped make Wyoming the center of coal production in the United States.

Coal is a cyclical business, and veteran coal men like Greg Boyce have spent decades riding out the busts and surfing the booms. Those who invest during bad times prosper when the tide returns. Demand climbs again, the price recovers, customers rediscover the appeal of cheap energy. Why should now be any different? That cyclical view has prevented the industry from grap-pling with, or even acknowledging, its current predicament. A future without coal, Vic Svec of Peabody told me, is a future of "soaring power prices, ex-ported jobs, and a languishing economy." The world can't do without coal.

At least that's how it's always been.

Two-term Wyoming governor Dave Freudenthal occupied the statehouse in Laramie from 2002 to 2010, a period that now seems like the last golden age for the coal industry. Freudenthal is a member of an increasingly rare species: a moderate Democrat who won two statewide elections in a state that strongly went for Barack Obama's opponents in the 2008 and 2012 presidential races. By nature a pragmatist and a technocrat, Freudenthal oversaw the passage of the nation's first state legislation to set standards and regulations for carbon capture and storage (CCS), the technology that many believe is the only path-way to survival for the coal industry. One reason Freudenthal connected with Wyoming voters is his plainspokenness.

"I think they are faced with a set of circumstances they're not equipped to really deal with," Freudenthal told me, referring to Greg Boyce, Peabody senior management, and the coal industry in general. "Some things you can control, and some things you gotta learn to accept. I don't think they had the wisdom to know the difference."

In Freudenthal's view, coal must accept a smaller role in the energy economy of the twenty-first century—at much reduced levels of production, consumption, and revenues, and only through technology that drastically reduces the carbon emissions from burning it. The governor has little patience for the complaints by Boyce, and other Big Coal executives, that the coal industry is being singled out as the villain in the climate change drama, that the so-called War on Coal is inherently unfair. "It doesn't matter how you feel about it," Freudenthal said, "some form of carbon capture is going to be necessary. They say, 'By God isn't fair.' Well, markets are not always fair."

The industry's current state, argued Freudenthal, stems from a "complete misreading of market expectations with regard to the provision of energy." Mark Northam, the director of the School of Energy Resources at the University of Wyoming, puts it another way: "The days of mine it, crush it, send it to a plant, burn it, and not worry about what comes out of [the] smokestack, are over," he told me.

Founded in 2006, the UW School of Energy Resources is charged with conducting research on ways of capturing Wyoming's abundant energy resources in sustainable ways. In many ways the school epitomizes the dilemma at the heart of the debate over Wyoming's future. Environmentalists dismiss the institute—which is supported by direct state funding as well as around $40 million in support, to date, from the private sector, mostly energy companies—as a tool of the industry. Many in the industry, meanwhile, see it as a vehicle for the development of new energy technologies that will supersede conventional coal-fired power generation. "When I go speak to the Rotary Club or some other group in Gillette, I'm too green," Northam chuckles. "When I go to Lander, I'm way too far on the side of the energy industry." It is past time, in Northam's estimation, for the coal industry to get realistic about the changing market. "You're never going to win the battle by saying climate change isn't real—the markets have already decided that question, and the industry has to realize that the future is going in a different direction. We're doing everything we can to help them make that transition."

For people like Greg Boyce, Scott Durgin, and the thousands of other Wyomingans whose livelihood depends, directly or indirectly, on unabated coal production and burning, that transition involves questioning the assumptions that have underlain their daily lives; acknowledging that their political leaders and corporate executives have been lying, or mistaken, for years; and generally reassessing their worldview. More than once in conversations in the basin, I heard reference to Kübler-Ross's stages of grief, as if the passing of the coal era were felt as a personal death. Acceptance is a stage that not many have been able to reach.

That is not to say that nothing has been done to envision, or prepare for, an alternative future. While the leasing of federal lands for coal mining may be a terrible deal for the taxpayers, the state since 1974 has collected severance taxes on mineral exploitation, revenue that goes into the Permanent Wyoming Mineral Trust Fund. As of 2013, the fund contained $5.6 billion.[36] At about $100,000 for every Wyoming resident, that's a pretty good severance package. How that money will be used is another question.

"There's probably another 25 years of mineable coal," Ann Turner, the publisher of the *Gillette News-Record*, told me in her office in the newspaper's building. "What happens then?"

Founded in 1904, the *News-Record* has been run by Turner's family for much of its history, and its views on energy production have mostly fallen in line with those of Gillette's coal workers. Turner herself spent six years living in Marin County, California, like a royal on a walkabout, before returning home to take up her designated role in life. But she does not talk like a mouthpiece for the industry.

"The coal companies should've recognized, a long time ago, that they needed to be partners in developing new technology," she said. "It's a little late now."

Turner, Northam, and Freudenthal all agree on the general outlines of Wyoming's preferred economic future: a gradual reduction in the production and burning of conventional coal; massive investment in R&D (research and development) for new, cleaner coal technologies; a more diversified economic base supported by a more skilled workforce; and a shift from the embattled, us-versus-them mentality that has thus far bolstered the aggressive pronouncements of Big Coal executives.

Wyoming residents aren't there yet. "We're in a state of denial," asserted Turner. "I don't know what it was like in Michigan, in the 1980s, when Japanese cars started appearing. But if we see a coal company go bankrupt, do you think that Obama would bend over backward to save it?"

Another resource for making the transition to the new energy age should be the Abandoned Mine Lands Trust, a federal program set up by Congress in the 1970s to compensate states for the environmental damage left behind by mining. Those payments, which reached into the dozens of millions of dollars a year, were curtailed in 2012, reducing Wyoming's annual payout from around $150 million to less than $14 million.[37] The funds were partially restored in September 2013, but the federal government still has nearly $3 billion in the Abandoned Mine Lands Trust, and no plan for disbursing that.[38] A good way to aggravate someone in Wyoming is to mention the Abandoned Mine Lands program.

The truth is that a shortage of money is not the problem. The will to face the future, and the imagination to envision it differently than the present, are what's lacking.

Some faltering attempts have already been made. First announced in 2008, the High Plains Gasification Advanced Technology Center was to be a state-of-the art research facility, attached to a power station, that would develop a process called coal gasification, in which physical coal is heated to a synthetic gas, or syngas, and then burned for power generation. The original project called for $50 million each from General Electric and from the state of Wyoming. GE pulled out in 2011, citing "uncertainty in the nation's energy policy," after the state had spent nearly $10 million on the launch.[39] Clean coal research continues at the School of Energy Resources, but the coal companies are little involved.

Before I left Gillette I visited the Dry Fork Power Station, a few miles north of town. Started in 2007 and opened in 2011, Dry Fork is a mine-mouth plant, meaning it operates entirely on coal from the adjacent Dry Fork mine. Rated at 385 megawatts—enough to power around 300,000 homes—Dry Fork, for a coal plant, is state of the art: it's zero discharge, which means all the water the facility uses is recycled within the plant, and it uses gigantic fans rather than water to cool the steam coming off the turbine that generates electricity. Dry Fork is among the least-polluting coal-fired plants ever built, manager Tom Stalcup told me proudly as he escorted me around the plant. The enclosed conveyor system that transports coal from the mine to the boiler eliminates the particulate blow-off from traditional coal piles, and the plant uses a "reflux circulating fluid bed dry scrubber," developed in Germany, to cleanse the smoke of sulfur and mercury.[40] Dry Fork is about as clean as conventional coal-fired power generation is ever going to get—and it will pump 3 million tons of carbon into the atmosphere over the next 40 years.

Dry Fork is not the future; it's a $1.35 billion remnant of the twentieth century, "a 1950s-era approach to generating power," as the Powder River Basin Resource Council puts it, most of whose electricity actually goes to other energy-extraction projects, i.e., coal mines.[41]

Mark Northam, of the School of Energy Resources, gave me a fairly bleak assessment of the prospects for Wyoming's coal industry over the next two decades. "The next five years are going to be tough, and they're going to be critical for the long-term survival of the industry."

Ten years from now, Northam sees some small commercial coal-to-liquid (and natural-gas-to-liquid) plants, either operational or under construction— but not enough to start a new coal boom. "I do not believe there will be a single CCS [carbon capture and sequestration] project in this region by that timeframe," he states, contradicting a half-dozen plans or projects being touted by the industry and the federal government.

Twenty years from now, he says, "will be a very different picture." Carbon capture and sequestration plants will be up and running, maybe on a commercial basis; coal gasification plants will produce syngas for electricity; coal will coexist with natural gas that is no longer so low-priced that other fuels are uncompetitive. After a decade or so of dwindling demand, a smaller, more sustainable coal industry will emerge. "I just hope the next several years are not so severe that coal companies start going under."

Or, maybe not. Maybe a more radical transformation awaits America's energy bread basket.

"Some people say that Montana, Wyoming, and the Dakotas should be turned back to the buffalo," Ann Turner told me, more rueful than bitter. "Maybe that's what'll happen."

Implausible as it seems, that's not an entirely fanciful scenario. More than a quarter-century ago, arguing that current uses of the arid plains are unsustainable, Rutgers professor Frank Popper and his wife Deborah, a geographer, proposed the Buffalo Commons: a 139,000 square-mile area, covering six Plains states, that would be returned to the buffalo. Already, wild, genetically "pure" bison from Yellowstone National Park have been transferred to open lands in Montana and Wyoming to restock the High Plains with their original inhabitants. (A gorgeous short film called *The Return,* available on YouTube, documents this migration.[42]) It's not clear whether the Buffalo Commons would be reopened to the Sioux, the Comanche, and the Arapaho, as well.

"In the East they're almost post-coal; here we're still in the middle of it," Shannon Anderson, of the Powder River Basin Resource Council, told me in our last conversation. "In Alaska"—where there are plans for a massive strip mine on the Chuit River, west of Anchorage—"they're looking at the future of coal.

"In terms of economic history, I think the past is our future: this whole region will go back to farms and ranches that use the land responsibly. We need to make sure that those farms and ranches have a future, once the mining is done, that they still have an aquifer, there's still grass to graze their cattle.

"That's really the question: Once the miners have left and the roughnecks are gone, what does the state really look like?"

On the Douglas Highway, heading south out of Gillette, the roughnecks and the coalfield suppliers, the machine shops and tire sellers and pipefitters that serve the energy industry line both sides of the road, petering out as it turns into a two-lane highway past Black Thunder and NARM and the other vast mines. From the road I could barely glimpse some of the open pits in the distance. They didn't look like scars. They looked like grayish clouds, lying low on the horizon, ephemeral as the weather. Then the road dipped, and they were gone.

CHAPTER 5

COLORADO

Jeff Troeger first came to Steamboat Springs in 1980, fresh out of grad school at the University of Indiana. Having grown up in South Bend and attended college in Bloomington, he'd always wanted to live out West. Back then Steamboat, set jewellike in the splendid Yampa River valley across two big mountain passes from Denver, was still a bit rough-hewn, a bit remote, and affordable for someone like Troeger. Plus, he and a friend had an idea for a business: they would sell analytics software to coal companies. The Yampa Valley was coal country.

"It was a good idea, but we soon realized that the coal market is on a one-year procurement cycle," Troeger told me. "We were way undercapitalized to sell to that market."

A brief and disastrous excursion into synfuels—another market whose time had not yet arrived—drained what little capital the fledgling company possessed. Living in a small cabin south of town, Troeger, not for the last time, shifted career gears. "So I started selling PCs: black market Apple 2s, Dells, DECs. I opened the first computer store on the Western Slope, in 1980 in Steamboat Springs."

Realizing where his true talents lay and foreseeing the computer revolution to come, Troeger soon began teaching basic computer science: spreadsheets, networks, databases. Eventually he joined the faculty of Colorado Mountain College, the network of two-year community colleges strung across the Western Slope, and taught there for 20 years. He married and raised two sons in the Yampa Valley. Having arrived in Steamboat before the real estate booms of the 1980s and 1990s, he was able to turn his ramshackle cabin into a house. And gradually, almost without realizing it, he became an environmental activist.

"I started reading Paul Hawken and Betsy Kolbert, and I started really looking around at what was happening here in the valley," he said. We were driving south out of Steamboat on a one-day tour of the area's energy history. Wiry and weathered, Troeger has a gray brush cut and dark piercing eyes. He gets animated when he talks about economics and the environment, and he gestured at the passing landscape: rolling green foothills dotted with stands of aspen and dense manzanita shrubs. Behind us rose the Steamboat Springs ski mountain, the trails yellow and green in early August.

"Basically we extract resources from the crust of the earth, and we heat, beat, and treat 'em to make temporary products that we then throw away in landfills, in the earth's crust. That's our economic system," he said.

Troeger's affinity for data, his concern for the local landscape, and his highly developed capacity for indignation all combined in his academic career when he began teaching sustainability studies at Colorado Mountain College in the mid-1990s. Later he helped launch the school's degree program in sustainability, the first such program in Colorado. At the same time, he was becoming deeply pessimistic about the future of industrial society and the consequences of climate change.

"I hate to say it but we need a crisis," he said as we wound up Highway 113, heading toward the Flattop Mountains. "There's no urgency, there's no sense of emergency. We can film a comet flying through space, we can detail the human genome. The coal companies use the 'TINA' argument: there is no alternative. I don't believe that. There are plenty of other sources of energy— it's way past time to stop burning rocks for our power.

"I don't want to stockpile tuna fish and bullets. That's not the world I want to live in."

Thinking about a global catastrophe, Troeger decided to act locally by becoming involved with the Yampa Valley Electric Association, the rural power co-op that supplies electricity to Steamboat and its neighboring communities, Hayden and Craig.

"I went to my first meeting of the YVEA and they literally met behind closed doors. I told the woman at the desk out front, 'I'd like to attend the meeting,' and she said, 'Do you have permission to attend?'"

"I had to swallow what I wanted to say: 'Wait a second. This is a public organization. Aren't there open-door laws? Aren't there sunshine laws you have to follow? Isn't this illegal?'"

Eventually allowed in, Troeger quickly realized that the YVEA, like many rural electric co-ops, is heavily coal-dependent. The association gets its power from Xcel Energy, the Minneapolis-based utility that serves eight states, including Colorado. The Hayden Power Station, one of Xcel's seven coal-fired power plants, sits on the Yampa River 25 miles west of Steamboat Springs. It was clear to Troeger and other progressives in Steamboat that YVEA, founded in 1940, needed to change its strategy and buy more clean energy. And it was also apparent that, under the board at that time, change was unlikely. "It became clear that the board was never ever gonna change. So we needed to change the board. We found candidates who were interested in adapting to the realities of energy in the twenty-first century."

Within three years the board had several new members, a majority of whom favored moving away from coal. Charting a new course would require the cooperation of the association's powerful, longtime CEO, Larry Covillo. Like many heads of rural power co-ops (also known as rural electric associations, or REAs), Covillo was entrenched, used to a compliant board, and resistant to change. "It's ironic," said Troeger. "REAs were formed during the Progressive era. But they fossilized over time and became conservative, hierarchical, and inward-looking."

In May 2012 the board met for the first time without Covillo present. Shortly after that meeting, Covillo was presented with a plan that included shifting the co-op's generation mix away from coal.

"Nothing like this had ever happened before," Troeger told me. "After all, it was Larry's 'lectric company! No one had ever dared tell him what to do."

At the October 2012 monthly meeting, Covillo surprised the board by announcing his retirement, effective in March 2013. He'd been at YVEA for 32 years. "I think he decided he had enough years for his retirement goals, could see which way the new wind was blowing, and decided he wasn't about to change," said Troeger.

Among the changes was an agreement with a Boulder-based company called Clean Energy Collective to build a community "solar garden" in the valley.

Headquartered in Carbondale, Colorado, Clean Energy Collective has helped pioneer the community solar model, in which individuals and businesses can buy panels in solar power generation facilities, rather than owning or leasing the solar panels themselves. Paul Spencer, the founder and CEO of the company, calls it "solar for the masses," noting that the vast majority of people either cannot afford solar power or do not live in places where adequate solar power is available. CEC signs a power purchase agreement with the incumbent utility, and then pre-sells solar generation capacity in the form of subscriptions (the company calls it buying solar panels) and finances construction using that income, essentially, as collateral. As the available panels get sold, the company pays off its lenders. Customers don't necessarily get the actual power flowing from the solar array; those electrons go onto the local power grid, and appear as renewable energy credits on the customers' bills. CEC makes money by charging customers a slight markup over the cost of producing the power. Utilities are happy because they receive credit toward state-mandated renewable energy standards; customers are happy because they get inexpensive solar power without the expense and hassle of putting panels on their roofs. As a way of shifting away from the antiquated, centralized, and coal-dependent power grid, community is a powerful model: CEC built its first solar garden in 2010 and now has 45 facilities spread across 19 utilities in 9 states. Spencer expects the number of facilities to double by the end of 2015.

In the Yampa Valley, though, CEC had a problem.

WHEN THE FIRST HUMAN BEINGS—likely ancestors of the Ute and Arapaho Indians—came over Rabbit Ears Pass and into the valley of the Yampa, they must have thought they'd found paradise. Most subsequent visitors and settlers have had the same response.

Rising in the mountains of southern Routt County, the Yampa flows placidly north, through Steamboat Springs, before bending sharply west and running across northwest Colorado and into the magnificent gorge of Dinosaur National Monument. Along most of its course the Yampa is a clear, shallow, slow-moving stream, marked occasionally by hot sulfur springs like the one that gave Steamboat its name (supposedly, French trappers mistook the sound of the spring bubbling to the surface for that of a paddle-wheeler). Before the arrival of Europeans, the Utes hunted the Yampa Valley in summer and partook of the medicinal qualities of the springs. Beaver ponds and oxbow lakes dot the stream's course, which in its undeveloped reaches is lined by dense growths of willow, elder, dogwood, and stately cottonwoods. Herons nest in the trees, flocks of migrating geese crowd the meadows, and kestrels and red-tailed hawks catch rising thermals high above the valley floor. Moose, bear, and elk are regular visitors. West of Steamboat the land changes abruptly, from lush grassland and subalpine hillsides to the red rock mesas of the high desert plateau. Cattlemen and farmers, arriving in the mid-nineteenth century, found the area as rich and alluring as had the natives, and the Yampa bottomlands quickly became home to some of the most prosperous ranchland in the Intermountain West.

The Yampa "is a new country," J. B. Dawson, the founder of what later became the Carpenter Ranch—900 acres of the most beautiful land on earth—wrote to his friends back east. "The public domain is all open and unfenced. . . . The hills out there are full of deer and elk and antelope, and the streams are full of trout."[1]

The arrival of the Denver Northwestern and Pacific Railway in 1908 connected the valley with the big cattle markets of the Plains, bringing a new level of prosperity to the isolated towns. And with the railway came the coal men. Settlers had burned coal in their stoves for years, and at the end of the nineteenth century James Harvey Crawford, the founder of Steamboat Springs, found a large deposit of coal just northwest of town. That field was later sold to the Elkhead Anthracite Coal Company, and more discoveries followed, near Craig and on the western side of the valley, in the shadow of the Flattop Mountains. By 1904 the Yampa coalfields had already been documented in the journal *Mines & Minerals,* and when the railroad arrived it found a ready market in Yampa Valley coal to be shipped east and south.

The Edna Mine was the first to open, followed by the Trapper Mine in 1977 and the Twentymile Mine in 1983. The paradox of the Yampa Valley is that for all the area's natural beauty, few of the skiers, fishermen, river rafters, and mountain bikers realize that Routt and Moffat counties form the heart of Colorado's large coal industry. Standing at the top of the gondola on Steamboat Mountain, looking across the valley, you are looking directly at the reclaimed Edna Mine, which closed in 1995. In August, when I visited, the old mine could be distinguished by its flattened ridge, covered in vegetation that's drier and less abundant than the surrounding, undisturbed slopes. Just beyond that ridge and to the west is the Twentymile Mine, owned by Peabody Energy.

Coal is also burned in the Yampa Valley. The Hayden Power Station feeds Xcel's power grid, while the Trapper Mine exists solely to feed the adjacent Craig Power Station, a 1,200-megawatt, three-boiler plant that supplies electricity to Tri-State Generation, which serves 44 electricity co-ops across Colorado, Nebraska, New Mexico, and Wyoming.

"Twentymile is the biggest customer of the power plants," Jeff Troeger said as we passed through the small town of Oak Creek and turned west, toward the mine. "They're sending power to the mines so they can mine coal to burn to make more power. It's a circular system."

Pull back the curtain of champagne powder ski slopes, the quaint downtown, the pristine valley, and thriving tourism economy, and what you find is coal. Craig, about 40 miles west of Steamboat in the mesa country of far west Colorado, has always been a coal town. And that made siting the new solar garden slightly problematic. Most of the prospective customers lived in Steamboat, at the eastern end of the valley. But land in Steamboat was not cheap, and CEC's business model was based, in part, on building solar arrays without paying too much for the land. Proximity to customers was a lesser concern.

"The YVEA covers Steamboat, it covers Hayden, and Craig," Paul Spencer told me. "And so naturally there was a lot of interest from customers to put it in Steamboat. But from our experience we knew that with its higher elevation, Steamboat gets ten times the snow Craig does. From a solar production perspective, we need to produce as much clean energy as we can, so we don't want this thing in Steamboat, we want it in Craig. Look at the dirt: it's a desert, there's a lot of sun there. That's where you want the solar array."

And there was an ideal site in Craig, on the north bank of the river, right by the sewage treatment plant and literally in the shadow of the Craig power station's smokestacks. Like the two Sevier plants in Tennessee, it was an apt juxtaposition of old energy and new. Spencer signed a power purchase agreement with the YVEA, a long-term lease, in exchange for free solar power, with the city of Craig and began signing up customers. CEC quickly landed enough customers to make the solar garden viable. That's when the problem arose.

The land the solar garden was on was owned by the city of Craig, but the underlying mineral rights were held by Tri-State, the operator of the Trapper Mine. The solar panels were planned for land that the coal company could, in theory, dig up at any time. Tri-State officials said the rights were unlikely to be exercised, but they declined to formally cede them. Spooked by the mineral rights issue, the title company on the land deal washed its hands of the arrangement. For a time, it appeared that new clean energy had been trumped, once again, by the interests of the old coal economy.

"It's been a disappointment for us and everyone involved, including Craig," Jonathan Moore, CEC's land manager, told the *Steamboat Pilot* newspaper.[2]

Paul Spencer and Terry Carwile, the mayor of Craig, weren't ready to give up. "We begged, borrowed, and stole," Spencer told me, chuckling. "We had to find a way to work around the mineral rights issue, and the town helped us do that."

"Yeah, there was resistance, some of it vocal," Carwile, himself a retired coal man, told me. "But I want that $1.2 million construction project to happen here in this community. It'll translate into economic benefits for Craig. It's not huge, but you gotta hit singles, not just home runs."

The opposition included local officials such as Moffat County commissioner John Kincaid, who has denied that carbon dioxide is a pollutant and who testified at an EPA hearing in Denver, in July 2014, that the rules limiting carbon emissions from power plants constitute a "War on Coal."[3]

In the Trapper Mine parking lot, the day I stopped by, a heavy-duty pickup truck sported a black bumper sticker that read, simply, "Coal. Guns. Freedom." Sometimes, though, the future sprouts from unlikely ground. Even some in Craig's coal community figured that clean, cheap, local power was probably a good thing. Spencer and Carwile, unlikely allies, kept moving forward. By the fall of 2014 a new, more amenable title company had been found, the deal was back in place, and CEC had resumed signing up customers. A truce had set in. In deepest coal country, small solar power was carving out a beachhead.

"Solar is not the replacement for coal," said Spencer. "That would take billions of acres. It's another power solution that helps build a low-carbon future.

We don't see it as a threat to the coal industry. In some small way, this project is an initial way to bridge the divide between Craig and Steamboat—between the coal-producing world and the renewable energies of the future."

IN TERRY CARWILE'S 30 YEARS working at the Trapper Mine, he ran just about every piece of machinery there is to run at a coal mine. "I ran loaders, dozers, scrapers, and the dragline. I was in line for a [full-time] dragline job but I was tired of rotating shifts. So I went back to the day shift. It was a monetary sacrifice but I was happy to do it."

He retired 30 years to the day after he started at the mine. I met Carwile at Downtown Books, the bookstore and coffee shop he owns in Craig. When I complimented him on the tidy storefront place, he said, "Thanks. Want to buy it?"

Built, literally, on top of coalfields, Craig has the tense, wary air of a besieged town waiting for the next assault to begin. Production at the Twentymile Mine was down 10 percent from the year before in 2013, reaching only 7.2 million tons, the lowest since 2001. The numbers have been going down since 2005, when production peaked at 9.4 million tons. The bottom has not been sighted. An oil and gas boom on the Western Slope briefly helped fill in the decline, but that's faltered too. "In '08 you couldn't find a motel room, with all the oil and gas activity going on," said Carwile. "Then the price at the Henry Hub fell through the floor, and all the oil companies fled. There were a couple of hotels built and we went from 40 percent under capacity to probably 40 percent over capacity."

Some ranches and farms still struggle to survive in the area, but water is increasingly scarce, and wheat, cattle, and sheep are hard livelihoods. Coal brings in nearly two-thirds of the tax revenue in Moffat County, and Carwile, like many of his constituents, is looking ahead at an uncertain future. "I understand why people are apprehensive for the future of Colorado extractive industries. I've seen six or seven mines go out of business since I've lived here. It's a finite resource: you've got X number of carbon molecules underground that are economic to recover, and after that it's gone. You can't protest or lobby Congress to put more coal in the ground."

As with many of the remaining mines on U.S. soil, estimates differ as to the amount of coal left underground at Twentymile; already, trucks travel several miles into the mine's tunnels to retrieve loads of coal from the long-wall and bring them to the surface. Peabody has a 16-year contract, which runs through 2028, to supply coal to the Hayden Power Station. Peabody says there's 34 million tons of recoverable coal left at Twentymile; in 2013 the mine shipped 7.2 million tons of coal. Coal company estimates of "recoverable reserves" are notoriously inflated, and some people estimate the real total could be closer to 20 million tons. Even taking Peabody's estimate at face value, simple math indicates that Twentymile will run out of coal before its contract with the Hayden plant runs out.

"There's about two years of recoverable coal left at Twentymile," asserted Jeff Troeger as we stopped along the highway to gaze at the white load-out at the mouth of the mine. A rail spur of the Union Pacific stretched in either direction. As of a couple of years ago, Peabody planned to fill that gap by starting production at the western end of the Wadge Seam, the same coal belt that's almost mined out at Twentymile. The new facility would be called the Sage Creek Portal, and there's an estimated 110 million tons of coal available there—if Peabody chooses to mine it.

In May 2012, Peabody announced it was moving forward with Sage Creek. Heavy drilling machines began carving out the underground tunnels, creating points of entry and establishing the faces where the coal would be mined. Little has happened since, though, and most locals have lost confidence that the project will move forward. Troeger and I drove up to the Sage Creek facility, past a disused security post. A few trucks were parked outside the low, modular office building. A U.S. flag flapped in the breeze, and a light burned over the door. But no one was about, and there was no activity on the dirt road up to the mine or at the mine site itself. Dark dug-up soil looked almost black against the green hillside.

"They've spent $200 million so far and now it looks like it's on hold," said Troeger. "I think Peabody is shopping that contract [with the Hayden station]. . . . I don't think Sage Creek will go operational, ever. I think they'll close the mine and walk away, and leave all this, defer the environmental cleanup costs. Hayden and Moffat County will suffer."

The same was happening across Colorado. The Elk Creek mine, near Paonia in southwest Colorado, shut down in late 2013 after an explosion set off an underground coal fire that couldn't be extinguished. More than 200 employees lost their jobs. A planned expansion of the neighboring West Elk Mine was halted in June 2014 by a judge who ruled that the U.S. Forest Service and Bureau of Land Management had approved the plan without considering the cost of the associated greenhouse gas emissions. It was the first time the courts ruled that climate effects had to be considered when granting permits for new mining operations in the United States. If that precedent takes hold it would effectively kill all new coal mining in the country.

Unlike Craig, though, Paonia has achieved something like a second life. Set in a gorgeous valley on the North Fork River, below the spectacular Grand Mesa, the town has become a tourist destination and a center for small-scale organic farming, with several renowned restaurants and a budding population of young arrivals who came for the land and the natural beauty, not the coal. Craig has few such amenities to fall back on.

"The Craig Power Station was really an economic game changer for this area," said Terry Carwile. "What are we going to do if that's gone? I don't know. . . . Things change. There's definitely change in the wind. You can get out front or get left behind. Either you're in the driver seat or you're getting dragged behind the bumper. I'm afraid sometimes it takes a big jolt to get people to jump in and take action."

That jolt could come in the form of new federal regulations. It could come in the form of Peabody finally washing its hands of Twentymile and the aborted Sage Creek project. Or, more insidiously, it could come in the form of attrition: the mines not hiring, letting the shovel and scraper and dragline operators retire, winding down as the coal plays out and demand declines. On Highway 40, called Victory Way as it passes through Craig, four o'clock traffic moved slowly through a sudden dust storm. The warning lights atop the smokestacks blinked slowly against the ominous sky.

BOB BEAUPREZ TOLD AN IMMIGRANT STORY to a small, restless crowd in Lincoln Park, outside the state capitol in Denver. It involved a man who came to the United States from Belgium at the turn of the twentieth century. He'd left his family behind in the Old Country to find work in the New. "Somehow he made his way to a place called Colorado," said Beauprez, his handsome features aflame in the fierce July sun. "He got a job shoveling coal in a power plant."[4]

The miner eventually saved enough to bring his family over to America. They settled in Colorado, near Lafayette, and made enough money to buy a dairy farm. Among his children was Beauprez's father, Joseph. "I'm here because of a coal job," said Beauprez. The crowd cheered.

Beauprez grew up in Lafayette, went to the University of Colorado in Boulder, went back to work the dairy farm for a few years, then sold it to developers to build a golf course. With the proceeds Beauprez bought a local bank and grew its assets from $4 million to $400 million in 12 years. A staunch free-market capitalist, he became the chairman of the state Republican Party and later a congressman. Now a candidate for governor, he was pumping up the crowd at a Friends of Coal rally, while inside the capitol the Environmental Protection Agency conducted a hearing—one of the only ones held in the western United States—on its proposed new limitations on carbon emissions from power plants. Beauprez was running against the Democratic technocrat, John Hickenlooper, who'd promoted clean energy initiatives at the expense of conventional fossil fuels. He was running against gun control, and higher taxes, and the leftward tilt that had seen Colorado become one of the first two states to legalize marijuana. And he was running against those who sought to shrink the coal industry.

"We have never used coal cleaner, safer, and more efficiently than we do right now," Beauprez declared in his anchorman's voice. "That isn't something to punish. It's something to celebrate. Why would you cripple the great American and the Colorado economy when it needs a helping hand the most?"

The crowd in front of the stage applauded on cue. A group of children holding up anti-EPA signs crowded behind the candidate. The organizers had handed out orange T-shirts with "Coal Keeps the Lights On" on the back and, on the front, an incongruously Soviet-looking logo of a flaming torch. Beauprez was the final speaker of a list that included Sharon Garcia, a Pueblo single mother and day care operator whose struggle to pay "skyrocketing" electricity bills (caused, in part, by rate hikes due to the local utility's transition from coal to natural gas) had been recently chronicled in the *Washington Post*.[5] The 200 or so sweaty attendees, almost all coal workers, were indignant, fearful, and bitter at being forgotten.

"They can hear you now!" shouted the emcee, a conservative radio show host, as the crowd did their best to roar.

Half the people there were from Craig, which had sent five busloads of miners, plant workers, and their families to the capital. "We asked the EPA to come hold hearings in Moffat County," Frank Moe, the owner of the Deer Park Best Western Inn in Craig and a newly elected county commissioner, told me. "They declined. How can you make an informed decision if you're not willing to travel to the places and talk to the people who are being most affected?"

I also met Sheri Herod, the wife of a coal miner and the mother of three sons, all of whom worked in the industry. She was fanning herself with a coal flyer. I told her I was coming to Craig in a couple of weeks. "Come see us," she said.

I saw her again, with her son Chris, at the Clarion Inn in Craig, just down the hill from the Trapper Mine. It was three o'clock in the afternoon. Chris was getting ready to start his shift at the mine. Sheri's husband, Jerry, was off that day, up at the lake. He'd worked at the mine since 1978 and was now one of the dragline operators, fourth in seniority at Trapper. At 58, he was getting close to retirement. "He wants to go on his sixtieth birthday," said Sheri. "It all depends on the market in the next couple of years."

Chris, 33, is the youngest of three brothers. "Dustin's the middle, Jason's the oldest. Dustin brings it out from underground, I take it through the processing plant, analyze the ash and moisture specs, wash it, ship it out on a train. Jason, he's a driver with Tri-State, out of Cortez right now. He transports it."

"Did you expect your kids to all work in the industry?" I asked Sheri.

"Noo-oh," she shook her head. "We wanted something better. We told 'em we don't want you to end up working in a coal mine, but that's what they all did. That's what we have, and they do well at it."

Jason went into the Navy after high school, got discharged and was working as a longshoreman in Seattle in September 2001. After 9/11 the work dried up, so he made his way back to Colorado and worked odd jobs till he got hired as an apprentice at Tri-State. Dustin, the second son, got a degree in computer animation from the Art Institute of Colorado, but he couldn't find a job in his field. He too was drawn back to Craig. "He married a girl with little boys and had to go to the mine and get the benefits," said Sheri. "The benefits are what everybody needs."

As for Chris, "All of my friends were working out at Twentymile, it seemed logical to me—I could waste money at college or go make a living. I went right into the workforce."

Chris had had a rough week. His best friend, another mineworker, had hit a deer on his motorcycle while riding home from work in the dark. He had a broken back and a broken pelvis. It was touch-and-go for a while, and they still didn't know if he'd recover fully. And there was the anxiety about the industry. Chris had two boys, seven and eight. The uncertainty gnawed at him.

When you ask people in Steamboat about the coal workers who are worried about the mines closing and the power plants shutting down, you usually get two reactions in succession: the first is sympathy, and the second, spoken or unspoken, is a question: Why don't they do something else? Why don't they move to North Dakota, where the industry can't hire enough rednecks and drivers and mechanics to fill the jobs in the gas fields? That's an easy question to answer if you come from a place and a part of society where mobility, both geographical and socioeconomic, is an accustomed part of life.

"I've never had a job for more than five years at a time," a prominent Democratic political strategist said to me in Denver, a few weeks after my Craig visit. "Nobody has a job for life anymore. If they have to move, that's part of the global economy that everybody lives in now." Unspoken was the political calculation: most coal miners vote Republican. If they leave the state they're not voting Republican in Colorado anymore.

"I came out of high school and went right into coal mining," said Chris Herod. "There's not much else I can fall back on. I believe our managers are trying to keep the mine going. I just have to have faith that they know what they're doing."

"You've got labor siding with management, even though the money, the real value of the coal, ends up in St. Louis," Jeff Troeger said with customary exasperation as we drove through the picturesque downtown of Hayden. "It doesn't stay in Craig. How do you split the two?"

I guessed that Troeger had a suggestion. I was right: "You've got to pay 'em to retire. What it would cost to retire the entire coal workforce, tomorrow?

Two billion dollars? Five billion dollars? A lot less than the bailout of Detroit. And this is way more critical to the future of the country."

BOB GREENLEE MOVED TO BOULDER from Iowa City not for the mountains, or the skiing, or the gorgeous coeds, or the world-class pot. He came for the radio stations. An ad executive, Greenlee had long nurtured a yearning for a more glamorous career in broadcasting. Not in front of the mic or the cameras, even though he has the senatorial looks and the plummy baritone to do that if he desired, but as an owner of radio stations. He wanted to live in a college town because college towns make good radio markets. In Boulder he found a quintessential college town and, better, a neglected daytime AM station for sale at a good price.

Buying an AM station in 1975 might have seemed like buying an aging coal plant now, but Greenlee had ambition, capital, and a strategy. In 1978 he went on the air with KBCO, an adult-oriented rock station that would become one of Colorado's most successful outlets and that would enable Greenlee to buy and run radio stations in Omaha, Tucson, and other medium-sized cities around the country. Before the advent of the Internet and of broadcasting giants like Clear Channel, a midsized radio chain like Greenlee's could thrive, and he made a small fortune. As successful businessmen often do, Greenlee then thought of getting into politics.

He ran unsuccessfully for the Boulder city council in 1981, but was appointed to a council seat a year later when one of the members died. He spent 18 years as a councilman and, in January 1997—ten days after the murder of the child pageant queen JonBenét Ramsey brought Boulder worldwide notoriety—he became mayor.

This was not a natural fit. Greenlee is a free market conservative in one of the most liberal towns in America. Often called the People's Republic of Boulder, the city has strict zoning laws, easygoing cops, a highly educated, wealthy population and a hip, freewheeling culture centered on outdoor recreation, self-realization, craft beer, and high real estate prices. Those lucky enough to live in Boulder, as I do, know they live in paradise, and they are extremely wary of anything that threatens their upscale bohemian gem of a town.

Greenlee, a moderate Republican, managed to get along with the more firebrand elements of the city leadership, and his term as mayor, though brief, was largely successful. He ran for the U.S. Congress against Mark Udall in 1998, lost again, and decided that his political career was over. He semiretired to oversee his business interests. But politics had not left him behind.

"The issue of where Boulder gets its power had been around, simmering in the background, for years," Greenlee recalled when I spoke to him in the summer of 2014. "Every time franchise renewal with the Public Service Company came up, every 20 years or so, people would wonder: 'Maybe there's a different way to get our electricity. Maybe we should consider who's going to supply our power, maybe entering into these 20-year extensions is not the best way to go.'"

In the early 2000s, Greenlee, having turned 60, had expected to be enjoying a leisurely life of tending his investments, running the Greenlee Family Foundation, and traveling with his wife. But those vague wonderings lingered and crystallized into a movement. Increasingly dissatisfied with the way Xcel Energy—the investor-owned utility that had absorbed the old Public Service Company that supplied northern Colorado's power—was plotting its energy future, the citizens of Boulder came to an audacious conclusion: we should own and operate our own municipal utility.

This would represent a return to an older era in the power sector, when many cities had their own generating stations and operated their own grids. Known as municipalization, such throwbacks had become a minor movement: 17 new public power utilities have formed in the last decade, according to the American Public Power Association, as cities have taken over their power supply from the incumbent utility, with varying degrees of success. Today there are 2,000 municipal utilities in the United States.[6]

The voices calling for municipalization in Boulder became increasingly loud as the evidence of catastrophic climate change mounted. Toward his fellow Boulderites Greenlee normally displays the bemused tolerance of a college professor for his more idealistic students. But when it became clear that municipalization might actually happen, he was aghast.

"It really became an issue within the last five years," he told me from his home in Lafayette, outside Boulder. "It was a combination of unhappiness with what seems to be going on in the utility industry itself, along with the extreme environmental concerns that the community of Boulder has. Attached to this is the climate change debate—in theory that's the main reason to municipalize, to rely on renewable rather than traditional forms of producing energy. In other words, to move completely away from coal. Boulder wants to be a pioneer in renewable energy: the stated goal is to have 100 percent renewables in the city of Boulder, which I think is a foolish concept."

Foolish not because it's not worth aspiring to; like most Boulderites, Greenlee believes that climate change is a threat and that moving off of coal is, in principle, a good idea. His custom-built home in Lafayette ("We finally couldn't live in Boulder proper anymore," he chuckled, only half kidding) has the largest solar installation allowable by law on a private residence. But Greenlee is a realist, and he thinks that Boulder going 100 percent renewable is a dangerous fantasy.

"Look, there's no question we're using nineteenth-century technology, developed in a twentieth-century regulatory environment, to supply our power in the twenty-first century. The utility industry has got to change. Right now we're going through the agony of how that change is going to work itself out. But when Boulder decided to go it alone—I just think that's a disaster waiting to happen, if they're actually successful in going that direction."

Greenlee, in other words, is a staunch adherent to the TINA school of thought: it would be nice to go 100 percent renewable, but the technology simply doesn't exist to do so today, or any time in the next 20 years, for that matter. The challenge of a small city like Boulder taking on its own power utility was certainly formidable. Municipalization, though, was like perpetual motion, or the gold standard: an idea that seized people's minds and made them dismiss the challenges as irrelevant.

The municipalization movement was led by PLAN-Boulder County, a civic group whose mission, in the eyes of the developer and business communities, seemed to be to raise the drawbridge, stifle healthy economic competition, and preserve Boulder in a hazy state of quasi-socialism, trust-fund elitism, and reggae jam-band music. Telling PLAN-Boulder that municipalization was too idealistic, too costly, and too futuristic was like showing a yellow card to an Italian soccer player. It only fed their determination.

Between 2004 and 2012 the city council commissioned multiple studies on the question of municipalization. A four-year study, completed in 2008, found that "although customer rates for electricity under a municipal electric utility have the potential to be lower, [they] may be significantly higher than Xcel Energy's projected rates." What's more, "significant legal and public relations challenges" to creating a municipal utility would represent millions of dollars of risk to the city.[7] Based on those conclusions, city manager Frank Bruno recommended that the plan be abandoned. Writing in the local newspaper, the *Daily Camera,* Greenlee rubbed his hands together with glee: "Every now and then Boulder's City Council does something so entirely uncharacteristic that one is astonished by its action. Pleasantly astonished this time because the council seems to have almost given up on its asinine desire to buy, own and operate the city's electrical distribution system."[8]

Asinine or not, the plan survived. Negotiations with Xcel over renewing the existing franchise agreement, which expired in 2010, grounded on the rocks of renewable power: Boulder wanted more than Xcel could or was willing to provide. Supporters of municipalization were driven by a simple and impassioned logic: the city had to decarbonize.

"We need to get rid of coal," declared council member Lisa Morzel. "We're not doing that with Xcel, and we don't have a real cooperative and willing partner."[9]

Finally, in November 2011, the issue went to the voters, who narrowly approved municipalization, with some conditions—including that electricity rates would be comparable to what people were already paying Xcel, that the municipal utility would supply reliable power, and that nearly all of that power would come from renewable sources. Over the objections of pragmatists like Greenlee as well as big companies that had local operations, including IBM, Boulder was going it alone.

In 2012 the city hired Heather Bailey, a former utility executive, as its executive director of energy strategy and electric utility development. At $250,000 a year (plus a housing stipend—this was Boulder, after all), she was the highest-paid city employee. She spent the first two years of her tenure dealing not with actually creating a power utility, but with fending off legal challenges to the plan. When the city council officially approved the formation of a Boulder-run utility, in May 2014, Xcel promptly sued the city, saying that Boulder officials had not demonstrated they could fulfill the charter requirements set forth by the voters.

Among other things, Xcel valued its stranded assets—the power lines and other infrastructure that already existed in the city and its surroundings, which the municipal utility would need to take over—at nearly half a billion dollars.[10]

By late 2014 the city was involved in five lawsuits and spending upward of $3 million a year, not to provide power but to pay outside consultants and lawyers to get ready to provide power. Critics like Bob Greenlee claimed the real costs were much more and that Bailey, and the city, were not being forthright about the magnitude of the challenge. Greenlee's own calculation of the total costs of municipalization, leaving aside the cost of Xcel's assets, was $1.2 billion.

"They have not fully disclosed to the public the real costs," Greenlee told me. "I think that Heather Bailey and her group, they know the things I know. But they are incapable of being honest and upfront about what they already know."

Even some city council members had qualms about the size of the challenge. "It's one thing to climb a mountain," said city councilman Ken Wilson in a public meeting. "It's another thing to jump off a cliff."[11]

THE FORCE THAT DROVE XCEL AND BOULDER to the edge of this cliff, the proximate cause of the municipalization battle, was coal. More specifically, it was the Comanche 3 coal-fired unit, near Pueblo, completed in 2010—almost certainly the last coal plant that will ever be built in Colorado.

Xcel officials point to Comanche 3 as the most advanced coal unit ever built—which, to environmentalists in 2014, is a bit like being the world's most advanced steam-powered locomotive. It's a perfection of obsolescence that cost $1.3 billion, and years of controversy and litigation, to build.

To be sure, Comanche 3 is a technological marvel. Its high-pressure "supercritical" boiler produces 5 to 6 percent more electricity per ton of coal than conventional units. Its airflow cooling units are projected to cut the total water consumed by the plant by half. Its state-of-the-art mercury-reduction systems could, one day, be retrofitted to existing plants. It's "carbon-capture-ready," which just means that if anyone ever invents a carbon capture system that will work, economically, at scale, it can be applied at Comanche. You can't burn coal much cleaner and more efficiently than Comanche 3 does. The most distinctive thing about Comanche 3, though, is not the machinery: it's the fact that stridently anti-coal environmental groups, including the Sierra Club, signed off on it.[12]

The plant arose as part of a historic compromise between environmental groups, which had sued Xcel over air pollution from its existing coal plants, including Comanche 1 and 2. A 2004 settlement between Xcel, the state public utilities commission (PUC), and environmental groups led by the Sierra Club decisively redirected Xcel's energy strategy toward cleaner forms of energy and away from coal. As part of the deal, Xcel agreed to install advanced pollution controls not only on Comanche 3, but also on units 1 and 2.[13] By generating new power at Comanche, the utility would schedule five other aging coal plants for closure. It would increase its share of power from renewable energy sources from around 2 percent in 2004 to 30 percent by 2020. Xcel also pledged to invest nearly $200 million in demand-side management programs, which lower demand by promoting energy conservation and efficiency among customers. Finally, and most strikingly, Xcel became one of the first U.S. utilities to factor into its resource planning models a projected future cost for carbon, through a tax or cap-and-trade or other system—a concession that infuriated some old-line officials, including Gregory Sopkin, then the chairman of the PUC. "The totality of evidence suggesting the inevitability of such a tax amounts to grim-faced witnesses declaring, to paraphrase, 'It's going to happen—you'll see,'" Sopkin told *Colorado Business* magazine.[14]

Environmentalists got a cleaner overall energy system in Colorado, they thought, and Xcel got its shiny new coal plant. The problem was the timing. Even as construction was winding up at Comanche 3, new drilling techniques like horizontal drilling and fracking were creating the shale gas revolution. The price of natural gas dropped to unforeseen levels, almost overnight, and suddenly a new $1.3 billion coal plant looked like a big, shiny white elephant—one that the company, and thus its ratepayers, would be paying off for decades.

David Eves, who became the CEO of Public Service Company, Xcel's Colorado subsidiary, in 2009, inherited Comanche 3. Boyish and affable at 56, Eves is hardly a reactionary utility executive. He has committed Xcel to leading the industry in terms of moving toward cleaner forms of energy. He would never say in public that Comanche 3 was a bad investment, but he's made it clear that Xcel has no plans to build more coal plants in the foreseeable future. Indeed, the irony of Boulder's municipalization fight is that the city is battling a company that, by the standards of the U.S. power sector, is a beacon of progressivism. Xcel even pledged to build Boulder its own wind farm that would supply 90 percent of the city's power by 2020. But for utility executives, even progressive ones like Eves, the coal calculus centers around the sunk cost fallacy: i.e., the wrongheaded notion that investment in prior years in existing assets determines future strategy concerning those assets. Aging, inefficient coal plants represent sunk costs that utilities are, however reluctantly, willing to write off. It's much harder to write off a new, barely depreciated plant that came online less than five years ago. Xcel and its customers are stuck with Comanche 3—a fact that tipped many Boulder residents against Xcel and for municipalization.

It has not helped matters that Comanche 3's supposedly ultramodern systems have been tough to keep in operation since the day the plant went online. Nor that Xcel went to the PUC for a $180 million rate hike to pay for investments in Colorado—in other words, to pay for Comanche 3. Xcel's high-handed approach to the municipalization battle—dismissing the effort at first as a puerile campaign spearheaded by lightweights and tree huggers, and later spending hundreds of thousands of dollars on an ill-conceived scare-tactic ad blitz, which largely backfired—has not helped its cause, either. Eves spent days in Boulder before a critical city council vote on the issue, lobbying council members in a series of backroom meetings that harked back to the days of closed-door sessions in smoke-filled rooms. In a clumsy covert maneuver, after voters approved municipalization, the company funded a local front group calling for yet another ballot measure to reverse the decision to municipalize.

"While it may be legal for Xcel to spend the profits gleaned from us on never-ending campaigns to defeat the will of our community, it is certainly unethical," wrote Susan Osborne, another former mayor, in an op-ed in the *Daily Camera*.[15]

It may be true that the People's Republic of Boulder would have decided to cut its ties to Xcel even if Comanche 3 had never been built. It's certainly true that, once Comanche 3 came online, the divorce became inevitable.

Comanche 3 was "a billion-dollar mistake," said Leslie Glustrom, a CU biochemist who has become the most outspoken advocate of municipalization in Boulder. "And now we're going to be paying for it for the next 30 years."

LESLIE GLUSTROM CAME TO BOULDER in 1992, seeking refuge. A born activist with a passion for taking on big industries, Glustrom had been living in Prescott, Arizona, for ten years with her husband, a teacher at Yavapai Community College, and raising their two kids. Drawn to controversy, Glustrom became outraged at the damage being done by privately owned cattle on public lands. She became an outspoken opponent of the Wise Use movement, which sought to preserve private grazing rights on federally owned open range. Taking on ranchers and the Bureau of Land Management in northern Arizona was a lonely struggle, and Glustrom waged it fiercely. As is her wont, she rubbed important people the wrong way.

Glustrom has a formidable recall for facts and figures, eyes the color of a glacier's inside, and a direct gaze that becomes unsettling when she gets on an extended tirade on a subject that excites her outrage, which happens often. She's a coal executive's worst nightmare: tireless, passionate, often better informed than the officials she's sparring with, and not shy about getting kicked out of public meetings, which had happened, most recently, a couple of weeks before we met in September 2014.

The mid-1990s saw the rise of the "range wars" across the Southwest, as activists concerned about erosion and overgrazing clashed with wealthy, often absentee ranchers and the federal bureaucrats who defended them. At the same time the militia movement was gathering force, as groups determined to protect their rights to stockpile firearms and use the land however they saw fit gathered force in places like northern Arizona. Leslie Glustrom was loud, visible, and quoted in the local papers. She began to draw hazardous attention.

"We'd go to a public meeting and the back third of the room would be militiamen wearing their hats," Glustrom told me. "They all had guns in their trucks. They knew where we lived. It was scary."

When her husband lost his job teaching English at the college, they decided it was time to decamp. They showed up in Boulder with little: "We had two kids, two mortgages, and no income." Glustrom, who had majored in biochemistry at the University of Wisconsin, walked into the lab of CU professor Deborah Wuttke and asked for a job, at any salary. She's worked there off and on ever since, with long breaks to work on her real vocation: shutting down the coal industry.

By this time, Glustrom had lost faith in the ability of governments and politicians to take meaningful action on climate change: "I realized there's nothing I can do about Congress; it's about the coal plants." Doing something about the coal plants has mostly taken the form of battling Xcel. Glustrom has been insulted, handcuffed, and given the bum's rush more times than she cares to remember. In April 2013 she was arrested, along with fellow Boulder activist Tom Asprey, while protesting at Peabody's annual shareholders' meeting in Gillette—a friendly site chosen, according to many observers, to avoid the level of protest that would have happened had the meeting been held at Peabody headquarters in St. Louis. Her standard mode is indignation, her default tactic confrontation. In 2011 she was actually banned from intervening further in Xcel-related cases before the PUC.[16]

She's also produced reams of reports and commentary refuting utility and coal company arguments, including a report on Comanche 3 (titled, naturally, "The Billion-Dollar Mistake") that claimed the new plant was essentially a moneymaking scheme: "There is strong reason to believe that the motivation for the coal plant was not to provide Colorado with the cleanest or cheapest energy solution, but rather to make a large capital investment so that Xcel could use the return on the capital investment for its shareholders and begin to recover from the stock price crash of 2002."[17]

When I sat down the first time with her, in the lobby of CU's airy and modern biochemistry building, she talked for half an hour before I asked a question. When I asked, "But don't you think that Xcel is—" she interrupted me, anticipating the question.

"Yes, certainly, Xcel is close to the top in the utility industry. The reason their commitment to renewables is as big as it is, is because we dragged them kicking and screaming, and now they're no longer kicking and screaming. They're reducing their risk—Eves realizes that sooner or later they must bear that risk of carbon having a price.

"The way I put it is, they're doing C work in a class of D and F students. It's not good enough. It's not good enough for the planet."

Glustrom dismissed self-described realists like Bob Greenlee as people unwilling to upset established institutions and dismantle existing power structures. "There's a million things we need to do, to go from a country that was nearly 100 percent dependent on fossil fuels, to get that as close to zero as we can. It's a daunting task. It's almost incomprehensible.

"We can't get there till we change the rules, 'cause we keep doing the same thing expecting the different result. We need a plan for getting Xcel's coal plants shut down by 2030 or earlier. Till we do that it's all just redecorating the staterooms on the *Titanic*."

To some degree, by Glustrom's reckoning, the coal wars have already entered their endgame; simply put, as in Craig, as in Harlan County, there's not enough coal that's profitable to mine to keep the furnaces burning.

"Look at Arch Coal's second-quarter earning report: in the Powder River Basin their profit margin is 18 cents a ton. You can earn more selling pencils on the street corner. What's happened in Kentucky and West Virginia is now happening in Wyoming. As soon as their production costs are higher than their profit margins, it's curtains. Xcel's working assumption is that coal falls out of the sky into the trains, and it will continue to do so for the next hundred years. That's not gonna happen."

Still, even though Boulder has, for the moment, won the municipalization fight, she is not certain of the outcome. In the middle of five legal battles with Xcel and other stakeholders, the city has spent at least $13 million so far without generating a kilowatt of electricity. Even among the most ardent Boulder greens, there are questions about the audacious plan to go it alone.

"It's a huge mountain to climb, under Colorado law, facing Xcel's connections," Glustrom told me. "There's no guarantee we'll succeed. Not everybody who climbs tall mountains gets to the top. The real goal is decarbonization. I'm confident that Boulder will continue on its path. Comanche 3 is supposed to operate till 2069. We'll decarbonize long before 2069."

To someone like Glustrom, who sees clear moral lines where others see fuzzy and ambiguous shapes, the logic of municipalization was obvious. To most Boulderites, it's less so. I got the feeling that plenty of my fellow citizens had voted for the plan not because they understood the challenges and benefits of municipalization, but because Xcel is a big, for-profit, coal-burning utility, and so by definition is on the wrong side of history. Without really intending to, they had taken sides in the struggle. That's what you do if you live in Boulder, whether the issue is prairie dog habitat or leash laws on open space or school funding: You take sides. You vote progressive. You topple the existing paradigm.

In this case the city may have taken on a struggle it ultimately cannot win. Xcel valued its sunk assets at more than a billion dollars, and demanded that the city pay a quarter of a billion to take them over; the city of Boulder, essentially, valued them at zero. Boulder filed a petition with federal regulators to seek a ruling on the issue. It was not clear whether the regulators, or the courts, would side with the city.

The latest report, commissioned by the city council and released in early 2013, found that the goals of municipalization—clean, reliable energy at a price at or below Xcel prices, with local control over the generation and transmission assets—were completely achievable, even if the city has to pay hundreds of millions for the stranded assets.[18] Reached by citizens' groups and consultants hired by the cities, these conclusions are dismissed by Greenlee and other opponents as deluded at best and deceptive at worst.

Ultimately the municipalization fight came down to evolution versus revolution: Can we shut down coal in time to limit catastrophic climate change by taking incremental steps, carried out under the existing power structure, or do we have to blow up the existing models and start over? The coal mines in Craig, the power stations of the Yampa Valley, the Comanche 3 coal-fired unit, the coal-bearing railways, all of the vast infrastructure built in the twentieth century to produce and transport and burn coal, all hang over this debate like a dead hand, as Upton Sinclair put it. Among Sinclair's novels was *King Coal*, a blistering exposé of the conditions faced by workers in the Rocky Mountain coal mines in the years prior to World War I. The "dead hand" was the term that Upton Sinclair gave to the economic, religious, and political structures that imprisoned and enslaved the mass of Americans in the early twentieth century. Municipalization means casting off the dead hand of the coal industry, but the big power generators have not only deep pockets, armies of lobbyists, and the TINA argument on their side; they also have the brute reality of an economic system built on cheap fuel—wrested from the land by laborers who are well paid and who have no desire to cast off their golden chains (to use a Marxist term)—and burned in plants that have taken years, and billions of dollars, to build and will take billions more to decommission and clean up. That's a lot of sunk costs. Boulder wants to be "a city on a hill," as one councilman put it, but even cities on hills need power lines to bring electricity to light their shining buildings. The advocates of municipalization had to acknowledge this.

"In its transition to 'the electric utility of the future,' it is unlikely that Boulder will be able to convert to 100 percent renewables on Day 1 of operations of a new municipal electric system," Boulder attorneys admitted in their petition to the Federal Energy Regulatory Commission. "Such a conversion would be phased in over time. During that phase-in period, Boulder will probably have to rely upon carbon-based generation for a portion of its power supply."[19]

David Eves claimed that Xcel was not resisting municipalization because of the precedent it would set for other cities considering setting up their own power supplies. But it was clear that if Boulder succeeded in breaking away, the whole rationale for multistate quasi-monopolies like Xcel would start to crumble. Clean, distributed, locally controlled power hastens the utility death spiral. Xcel was willing to spend nearly unlimited amounts of money to prevent that from happening.

Meanwhile, Glustrom, PLAN-Boulder, and the city council members who voted for the plan placed the community's right to choose carbon-free energy above all other considerations of price and practicality. The risks of failure were nothing compared to the risks of inaction. They had come too far to think otherwise.

"To do that would mean, not merely to be defeated, but to acknowledge defeat," Sinclair wrote, more than a century earlier, "and the difference between these two things is what keeps the world going."[20]

PART III

THE GREAT MIGRATION

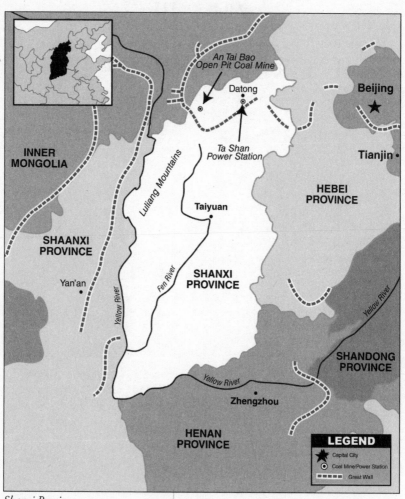

Shanxi Province

CHAPTER 6

SHANGHAI

The *Pu Jiang Youian* pushed slowly away from the quay and puttered into the stream. The Huangpu River, bisecting Shanghai, was crowded with watercraft; luxury yachts, workmen's scows, barges with cargoes under tarpaulins, and other pleasure boats like the one we were on slid past in the slow-moving current. Decked with holiday lights and a glowing neon sign, our boat was a floating, two-level party; attendees queued for the buffet dinner, dined at big round tables, milled about on the open-air upper deck, and admired the view. A string quartet of dark-haired young women in black cocktail dresses played Mozart. The evening had the soft fragrance of a spring night in a great Asian city.

We passed under the windows of the world's most famous riverfront, the Bund, China's financial center since the early eighteenth century. After World War II, as Mao tightened the communist garrote around the country, foreign traders and bankers fled the Bund and the stately buildings along the waterfront suffered. Now, in 2014, China's economic miracle was in its fourth decade, the financial industry had returned in force, and the investment houses and banks and trading firms along the Huangpu lit the night with a renewed dazzle. The same month of my arrival, the International Monetary Fund reported that by at least one measure—the purchasing power of the yuan, China's currency, versus the dollar—China's economy was set to surpass the United States before the end of the year.[1]

Across the river, on the eastern shore, the futuristic skyline of Pudong glittered like a collection of fantastic toys. Like dowagers eyeing the frippery of youth, the nineteenth-century buildings of the Bund presented a stolid face to the flamboyant skyscrapers across the river.

On board the *Pu Jiang Youian* were the leaders of China's coal industry: mining tycoons, government officials, expatriates from major coal exporters, investors, and a handful of trade-journal editors. This evening was the welcome dinner for Coaltrans Shanghai, the major annual coal conference in China. The free dinner, the flowing wine and beer, the classical music, and the chatter of the partygoers masked a growing sense of unease. China's coal industry, which during the boom years had become by far the largest in the world, was headed for major disruption that the revelers couldn't ignore.

China became the world's largest producer and consumer of coal in the 1990s, and by 2012 it was also the second-largest importer of coal, behind Japan. China has the third-largest reserves of coal in the world, behind the United States and Russia, and it burns about as much coal every year as the rest of the world combined.[2] Unlike the United States, which even at the peak of the coal era had a relatively diverse energy base, China's economy is heavily coal dependent: the country gets more than three-quarters of its electricity, and nearly 70 percent of its overall energy, from coal.[3] China's economic miracle was unachievable without coal. You can't talk about the future of energy without talking about coal, and you can't talk about coal without talking about China.

But in the last decade this cheap energy source has exacted costs that are no longer supportable. Air pollution now chokes the major cities and envelops the countryside; according to the Global Burden of Disease study, produced by Harvard University and seven other partners and first published in the medical journal *The Lancet,* 1.2 million people die prematurely every year from air pollution in China.[4] That's about the population of Dallas, dying every year, mostly because of coal. Coal trucks choke the major roads, and huge coal plants litter the countryside and the outskirts of every city. Each year the country's coal plants produce 375 million tons of toxic coal ash—enough to fill an Olympic-sized swimming pool every two and a half minutes, according to Greenpeace.[5] The huge mines in the north and northwest parts of the country periodically spawn enormous storms of coal dust that tend to follow the prevailing winds, toward the big cities of the coast. Although safety in coal mines has improved in the last ten years, the fatality rate in the mines is still appalling: nearly a thousand miners died on the job in 2013. The government spends more than $1.2 billion a year to subsidize the consumption of coal.[6] Coal in China is ubiquitous, dirty, and deadly.

To say nothing of its effects on the rest of the world. Now the world's largest emitter of greenhouse gases, China will see its emissions soar in the next 40 years unless the coal industry is dramatically curtailed. By 2030, if current trends continue, China could be putting out 10.7 billion tons of carbon a year from coal—nearly as much as the country's total output from all sources in 2012.[7]

In short, it doesn't matter what the rest of the world does about carbon pollution; unless China weans itself from coal in the next two decades, there is no chance of limiting global climate change.

On November 12, 2014, President Barack Obama and Chinese President Xi Jinping reached a historic bilateral agreement to reduce their countries' production of greenhouse gases. Calling for the United States to cut its emissions by 26 to 28 percent of 2005 levels by 2025 and for China to cap its emissions by 2030, it was a landmark deal between the world's two largest sources of carbon dioxide. But it was entirely voluntary, and reaching that goal would require China to take drastic steps to rein in the burning of coal—steps that could slow economic growth and sow unrest across the country. A week after the Obama-Xi handshake, the central government said it would cap coal consumption at 4.2 billion tons a year and reduce coal's proportion of total energy to 62 percent by 2020.[8] The problem was that it's not certain that those measures will be enough to reduce carbon emissions after 2030—and it's not at all clear that they will be carried out.

As in the United States, though, by the time I arrived in Shanghai, market forces had already begun the job of shrinking the coal industry. China's construction boom has slowed, reducing demand for steel, and thus for coal, even as the government pours billions of dollars into creating less damaging forms of energy generation—nuclear power, natural gas-fired plants, solar and wind farms, huge hydroelectric projects on the rivers of the interior. The worldwide price of coal has plummeted, making many of China's coal mines unprofitable even as the government attempts to shut down smaller, unlicensed mines. To limit overproduction and reduce excess supply in the system, the government is reportedly considering instituting absolute coal production caps in the next year. Its ability to control the output of the big mines in the interior, however, is limited.

The coal industry in China is broken, but, like a party barge on the Huangpu, it still runs, chugging relentlessly upstream, a vast state-within-the-state with its own cities, its own security forces, its own economics, and its own inertia. "For the foreseeable future, coal will be irreplaceable in China's energy supply," said Qu Jianwu, director general of the Shanxi Coal Transfer Center, during his address at the Coaltrans conference.[9]

Shutting down coal in China is literally a matter of life and death for hundreds of millions of Chinese and, quite possibly, for billions of people who live in other countries. "China's coal consumption has become the single most significant determinant for the future of the world's climate," wrote Greenpeace in an April 2014 report.[10] On the central government's efforts to limit the burning of coal rests the future of the earth's environment. In Shanghai, during the coal industry's biggest annual gathering, this was not an encouraging prospect.

COAL HAS BEEN INTERTWINED with Chinese society since the earliest days of the empire. In the northern and western provinces, where trees for fuel were scarce and coal seams could be found close to the surface, there is evidence people were digging and burning coal for household use as far back as the fourth millennium BC—centuries before the societies of the West discovered coal as a fuel. By 2000 BC, in present-day Inner Mongolia and Shanxi Province, coal was being mined and used to heat dwellings and to smelt copper.

With the coming of imperial China, the mining and use of coal became systematic across much of the empire. One reason China was first in so many artistic and technological accomplishments was because of the cheap energy that coal made available. Many of the achievements of the later Han dynasty (206 BC to 220 AD)—elaborate lacquerware, exquisite bronze work, the perfection of the papermaking process, and the development of the wind-powered bellows—were made possible by this energy surplus.[11] While the Renaissance was still three centuries off in Europe, coal enabled the emergence of a sophisticated, centrally governed, technologically advanced society in China. "When China began using coal to make cheap iron in the eleventh century, . . . coal and iron spurred industrial development on a scale that the world had never before seen, and would not see again until Britain's industrial revolution," writes Barbara Freese in her anthropological history of coal, *Coal: A Human History.*[12] Abundant energy from coal allowed urbanization on an unprecedented scale: Kaifeng, the capital of the Northern Song dynasty (960–1125), had as many as a million inhabitants at a time when Paris had fewer than 200,000 souls and London was a muddy settlement of barely 20,000.

By the early twentieth century, the coal industry, though still fragmented, extended across China. The production and use of coal expanded after the forced opening of the empire following the Opium Wars of the mid-nineteenth century. Despite the chaos that engulfed the countryside and the withering of the central government in the last decades of the Qing dynasty, coal helped accelerate industrialization: the first railway line officially sanctioned by the government was built to reach the Kaiping coal mines in Hebei Province. Coal fueled the railway explosion, funded by European banks, that saw 1,000 miles of new rail lines laid between 1912 and 1920, stretching from Canton in the south to Manchuria in the north. China's first large-scale industrial complex, producing iron and steel, was built in the early years of the twentieth century at the Han-Ye-Ping coal mines in Hubei Province.[13]

Coal also played a central role in China's revolutionary politics. During the Battle of Shanghai, the National Revolutionary Army fought to hold off imperial Japanese troops in one of the major engagements of the Second Sino-Japanese War. In the three-month siege—which ended with the fall of the city on November 26, 1937—the burgeoning Shanghai coal industry mobilized to aid the soldiers, forming a battlefield rescue operation called the Shanghai Coal Industry Ambulance Corps. Coal laborers were a major source of recruits for the Communist Party, and after the fall of Shanghai dozens of the drivers, along with their vehicles, joined the New Fourth Army, the fledgling outfit that would become a key unit of the communist Chinese forces. The coal workers' support for the revolution "led to a triangular relationship between coal industry capitalists, the [ambulance] corps itself, and the Communist army based in the countryside," writes Allison Rottman in her history of revolutionary Shanghai.[14] That relationship helped to shape the complex dynamics of wartime China as well as the history of Maoist China after World War II. Like Stalin, who gained fame organizing the oil workers in Baku, Mao got his start in the energy industry. At the Han-Ye-Ping complex, where more than 20,000 workers toiled, Mao cut his teeth as a labor organizer, helping to organize the famous 1922 strike at the Anyuan mine, which served as a springboard for the revolution.[15]

China's coal wealth formed a key piece of the vast mosaic of conflict that would coalesce into World War II: energy-poor Japan invaded northern China in the 1930s in part to seize the rich coal deposits of Manchuria.

After the war, energy from coal helped fuel Mao's disastrous economic experiments—although coal shortages became a major factor in the disastrous attempts at industrial development (including the notorious "backyard steel furnaces" that produced useless pig iron rather than steel) during the Great Leap Forward. Big state-owned mines were established under central planning, becoming centers not only of industry but housing, education, and services—the "coal cities" that would dominate economic life in many provinces well into the twenty-first century.

Once economic reform took hold in the 1980s, the country's hunger for coal to produce steel, concrete, and electricity became insatiable. The architects of reform under paramount leader Deng Xiaoping realized early on that economic growth depended entirely on abundant supplies of coal. They adopted a policy of unfettered development of large, medium-sized, and small mines simultaneously. If you had a pickax, a mule, and a cart to load, you could become a coal miner. Millions did.

In 1987 coal production surpassed 1 billion tons for the first time. From 1980 to 1996, production more than doubled, from 683 million tons to more than 1.4 billion.[16] In the coal-producing regions—the coal belt that stretches from Inner Mongolia to Xinjiang in the far west, encompassing the whole of Shanxi Province in the north—mines sprang up with little regard for safety, the availability of water, transport capability, or environmental consequences. In truth, by the turn of the twenty-first century the Chinese coal industry had changed little since before the war. Mechanization had come to the big mines, but in most mines the work was done mostly by hand, performed by men willing to crawl into dark underground chambers and hack away at the rock face because they had few other options for making a living. Profits were as scarce as safety: coal industry losses before the reforms of the 1990s stretched into the hundreds of millions of dollars per year. The five-year plan for 1991–95 called for the elimination of 400,000 coal mining jobs, nearly 6 percent of the 7 million workers in the industry.[17] Prime Minister Zhu Rongji, known as "One-Chop Zhu" for his ruthless management style, was shown on state TV in 1992 angrily lecturing coal managers at a large mine in Shanxi Province for their waste and profligate hiring practices.[18] Slowly, the industry was forced to modernize.

At the same time, the State Power Corporation—the state-controlled monopoly that generated and transmitted electricity in China up through the reforms of the mid-1990s—launched an astonishing binge of power plant construction that continued through the first decade of the twentieth century.

After the revolution, in which the communists took control of the shattered country after World War II and the civil war that followed, the power sector in China was small, fragmented, and outdated. Total national capacity was only 1.85 gigawatts—a small fraction of the capacity of, for example, the state of California at the time.[19] Over the next seven decades the country embarked on the greatest and fastest expansion of power generation the world has ever seen. Energy use per capita was far below the world average, to say nothing of developed countries like the United States; but it began to climb steadily in the 1980s, finally matching the world average around 2008.[20] All of Asia Pacific accounted for only about 15 percent of total world energy consumption in 1971; by 2010 that figure had ballooned to 38 percent, driven mostly by growth in China.[21] Total electricity use in China in 1980 was 250 terawatt-hours, only slightly more than the state of California. By 2010 that number had ballooned to nearly 4,000 terawatt-hours—almost as much as the entire United States consumed.[22]

Most of that new power came from coal. During the eighth five-year plan, which ran from 1991 to 1995, total coal production in China grew by 40 million tons a year. Between 1997 and 2005, China added 206 gigawatts of power generation capacity, or 500 megawatts—the equivalent of a medium-sized coal-fired power plant—*every week*.[23] These are official figures; the actual total is probably more, as illicit mines proliferated and many companies set up their own private coal boilers, off the national grid, to run factories and steel mills and cement plants.

By the turn of the twenty-first century, the central government had begun to realize that the country's dependence on coal was a devil's bargain. With annual production growing at nearly 10 percent a year, officials in Beijing began a program of closing small mines that managed to reduce production from just under 1.4 billion tons in 1996 to less than 1 billion in 2000. That proved to be a pause: the worldwide construction boom of the 2000s fueled a seemingly limitless expansion of coal. Coal production doubled from 2001 to 2004, reaching 2 billion tons; by 2009 it was 3 billion, still not enough.[24] Imports climbed as well, and in 2007 China, for the first time, became a net importer of coal.[25]

That same year, China surpassed the United States to become the world's largest emitter of carbon dioxide.

CHINA IS STILL GROWING at 7 to 8 percent a year, a rate that all of the Western industrial nations envy. But the model is faltering, and there are increasing signals that a powerful and unpredictable transition is gathering force.

"We have hit the limit of this type of growth," Zou Ji, deputy director of China's National Centre for Climate Change Strategy, told the online news outlet *China Dialogue* in 2013.[26]

In June 2014, the energy information provider Platts reported that Shenhua Group, the state-owned coal giant, was pleading with utilities and coal traders to reduce oversupplies at Shenhua loading facilities.[27] For the world's largest coal producer, this was an extraordinary development. Created in 1995 by the State Council of the People's Republic, Shenhua had grown into a vertically integrated behemoth with interests in not only coal mines, but also railways, power plants, ports, shipping, and coal liquefaction. Its revenue had grown in tandem with China's booming economy, reaching more than $46 billion in 2013, nearly twice as much as those of Exelon, the largest U.S. utility in terms of revenue.[28] Enjoying deep connections with the Communist Party leadership and the People's Liberation Army, it epitomized the for-profit state enterprises that dominated Chinese economic life. Even as the coal boom began to slow, Shenhua's low-cost mines and its sheer scale buffered it from market forces that battered other coal companies: its coal sales grew by more than 10 percent in 2013. In 2012, with its fleet of coal-fired plants growing across the country and its thirst for coal growing faster than its domestic production could handle, Shenhua had announced that it would spend more than $600 million on new coal loading terminals on the coast.[29] Unable to satisfy demand with its domestic reserves, Shenhua was looking abroad for coal, planning a giant strip mine in New South Wales, Australia, that was expected to produce 11 million tons a year for 30 years, and entering into a $10 billion joint venture with Russian company Rostec to develop the vast coal resources of Siberia.[30] The coal market might be in transition, but Shenhua seemed untouchable—until 2014.

Now, unsold coal was piling up. Coal stocks at the port of Huanghua, on the northern coast, had jumped 68 percent in two months, reaching 2.4 million tons. It was costing Shenhua millions of yuan a day to have the coal just sit, and now it was pleading with its customers to take the stuff off its hands. Shenhua's net profit fell by nearly 10 percent in 2013, which was still better than its competitors: China Coal earnings dropped 57 percent for the year, while Yanzhou Coal plummeted by 87 percent. China's coal mining sector, for the first time in decades, was facing a major contraction. Perhaps 40 percent of the coal producers in the country lost money in 2013.[31]

China's coal glut helped cause a disastrous drop in world prices: spot prices at Newcastle, Australia, the world's largest coal export terminal, had fallen by more than half, from $142 a metric ton in January 2011 to $62 in October 2014.[32] Still, defying the laws of supply and demand, the country's coal production edged inexorably upward in 2013, to 4.1 billion tons, according to the China National Coal Association.[33]

A thousand miles west of Huanghua port, in Inner Mongolia, the industry's precipitous decline had returned once-bustling towns to their pre-coal torpor. The Baofu highway, a primary transit route from the mining center at Ordos, was so crowded with coal trucks a few years before that traffic jams often lasted for days. Now, according to Reuters, the road was largely empty. "Rows of once busy restaurants are closed and flanked by advertisements for discounted coal," Reuters reported. "At mines that are still operating, unsold coal is piled high and lacking its black sheen, having been exposed to the elements for months."[34]

News outlets on the mainland were reporting pay cuts of up to 50 percent at big mining facilities, while many smaller mines were simply shutting down. Like their U.S. counterparts, China's coal miners have been battered by a combination of market forces and government policy. The market side of that equation is simple: China's overheated economy is cooling as demand for steel falls in China and abroad, the country's construction and infrastructure-building binge slows, and the government seeks to move toward a more sustainable economic model based more on domestic markets and less on cheap exports. China's GDP grew at 7.7 percent in 2013, matching 2012 for the lowest rate in a decade.[35] Slower economic growth means lower demand for coal.

In March 2014 Shanghai Chaori Solar Energy became the first domestic Chinese company to default on its corporate debt, prompting alarmed talk of a possible wave of defaults. Weeks later came the next default, this one much closer to the coal industry: Haixin Steel, a major steel producer based in the coal heartland of Shanxi Province, missed payments on 20 billion yuan ($3.21 billion) in bank loans, sending a ripple of alarm through the steel sector, whose furnaces provide a major customer for China's metallurgical ("met") coal. The Beijing-based investigative journal *Caixin,* a primary source for information on corruption and insider trading among China's business and government elite, reported that Haixin's debts are actually as much as seven times larger than officially acknowledged. In fact, the entire steel sector could be facing a debt crisis: steel producers borrowed 1.5 trillion yuan ($241 billion) in 2013, said *Caixin,* most of which is unlikely to be paid back.[36] Cheap money has long kept unprofitable companies afloat in the steel, construction, and coal industries—an era of easy lending that is coming to an end.

At the same time, the structure of China's economy is shifting. As modern economies grow more sophisticated, their energy intensity—the amount of energy required to produce one unit of GDP—tends to fall. To economists, this is the environmental version of what's known as the Kuznets Curve. Energy intensity, and the accompanying pollution, in the West rose for nearly a century after the beginning of the Industrial Revolution, then started to fall as technology improved, products became more inventive, and people demanded cleaner water and clearer skies.

China shows signs of following that curve. It takes less coal to assemble an iPhone than it does to make an airplane.

China's central government, meanwhile, is leading a "War on Coal" that makes the Obama administration's efforts look like a Quaker meeting. The first targets are the small, independent, often unlicensed mines that lace the countryside of the northern and western provinces. Most of these mines are outdated, underground, dark, and dangerous: by some estimates up to 90 percent of China's coal mining deaths occur in small mines. The campaign to eliminate illicit and unprofitable mines has been going on for more than a decade, but it has gathered force in recent years. In April 2014 the National Energy Administration announced that it would close 1,725 small-scale mines by the end of the year, with a total capacity of 130 million tons a year—less than 4 percent of the country's total output, but a significant fraction nevertheless.[37]

Thousands more of these old and nearly depleted mines will go out of business in the next decade, to be replaced by larger mines that produce 100 million tons a year or more. Many of these big mines are in the North and the northwest, assembled in vast "coal bases" that I will describe in detail in chapter 8. The Small Plant Closure Program, launched under the eleventh five-year plan (2006–10), seeks to do the same thing on the power generation side, eliminating aging, inefficient coal burners and replacing them, not with nuclear or renewable sources, unfortunately, but with newer and more efficient coal burners.

At the same time the People's Republic of China is inching toward absolute caps on both coal consumption and carbon emissions—moves that could spell the beginning of the end of the coal era in China.

Reducing the consumption of coal to 4.2 billion tons by 2020, as called for in the announcement following the carbon reduction deal with the United States, doesn't sound like a huge achievement. But it could cause a historic contraction in the coal industry and eliminate thousands of jobs in mining, processing, and burning coal. The central government has also instituted hard limits on the burning of coal in three regions: Guangdong Province, the industrial area around Hong Kong and Shenzhen, in the south; Shanghai and its sprawling environs; and Hebei Province, which encompasses the Beijing and Tianjin municipalities. These, of course, are the major centers of commerce and foreign investment in China, and it is imperative to clean up the air above them.

"Furthermore, many provinces are now committing to reverse the trend of rapid growth in coal use and cut their coal consumption overall in just four years," said an April 2014 report from Greenpeace, "The End of China's Coal

Boom." "No other major coal consuming country has ever implemented such rapid changes in their coal policies."[38]

So far 12 of China's 34 provinces have pledged to implement absolute coal consumption targets. The financial community has also realized that coal in China has passed its prime. Recent analysis by Goldman Sachs concluded that miners now lose 15 percent on every ton of coal they sell.[39] An influential report published in September 2013 by Citibank, titled "The Unimaginable: Peak Coal in China," marked the first statement from a major financial institution that the growth in coal use in China is not limitless. The combination of government curbs and market forces will cause coal use to flatten and begin to fall much earlier than most experts have predicted, argued the Citibank analysts. "Although many global energy agencies continue to expect high coal demand in the years to come," they wrote, "Citi expects this combination of factors"—clean-air programs, structural shifts in the economy and energy intensity, and robust growth in renewable energy and nuclear power—"to slow the power sector's use of coal, pointing to a possible flattening or peaking before 2020." The effects will ripple across the global economy: "Although lower prices may spur demand growth [for coal] elsewhere, the demand slowdown in China should more than offset such gains, in our view. Coal exporting countries that have been counting on strong future coal demand could be most at risk."[40]

THAT WOULD BE VERY, VERY good news for the world's climate. If these reductions are achieved, by 2020, 665 million fewer tons of coal will be burned, saving 1.3 billion tons of carbon dioxide emissions, according to the Greenpeace report—an amount equal to total emissions from Canada and Australia combined. A reduction on that scale would bring China's emissions into a range not far off from the levels needed to help keep global warming to 2 degrees Celsius. It "will not only fundamentally shift the coal consumption trajectory of the world's largest coal consumer, but also significantly re-shape the global CO_2 emission landscape."[41]

As always in China, however, there is a counterargument, a countervailing force. Even as it took steps to rein in illicit production and move dirty industries, including power generation from coal, away from the big cities of the coast, the government was approving one of the biggest coal mine construction programs in history. In 2013, planners in Beijing approved more than 112 million tons of new coal production capacity, according to Reuters. If all the planned new production actually comes online, the country will have added 860 million tons between 2011 and 2015—more than India produces in a year.[42] At least 15 big new mines will open in the coming years, all in the new coal bases in the north and west.

China's energy use, meanwhile, is expected to more than double from 2010 to 2040, reaching 220 quadrillion BTU.[43] That power has to come from somewhere. A lot of it is going to come from coal.

Power generation from coal will still account for 64 percent of China's supply in 2030, close to the current figure, according to consultancy Wood Mackenzie, which has offices in Singapore and Beijing and which dismisses the encouraging declines in coal use in the coastal provinces. "China's coal story," says Gavin Thompson, chief of Asia Pacific gas and power research for Wood Mackenzie, "is far from over."[44]

In this scenario, hopes that China's coal consumption will begin to decline around 2020 are wishful fantasies. "It is very unlikely that demand for thermal coal in China will peak before 2030," said William Durbin, the Beijing-based president of global markets for Wood Mackenzie, in a statement accompanying the release of a report entitled "China: The Illusion of Peak Coal." "Despite efforts to limit coal consumption and seek alternative fuel options, China's strong appetite for thermal coal will lead to a doubling of demand by 2030," the report concluded.[45]

Nevertheless, the sense in the industry that the boom times were over was strong. Normally, an industry conference like Coaltrans Shanghai would be an opportunity for some glowing speeches, a few self-congratulatory panel discussions, plenty of booze-fueled deal-making, and a couple of late nights of carousing. This one, held amid the turmoil of China's coal industry in 2014, was downright gloomy.

"The world is unstable," pronounced Dr. Yang Yufeng of the government's Energy Research Institute. "The economy is moving from a decade of high growth into a period of retrenchment and regeneration. The energy industry is facing many uncertainties. The whole global pattern of energy is changing, thanks to geopolitical and economic issues. The coal industry in China has not responded to these changes."[46]

Reducing pollution by reducing coal burning, Yang declared, is not just a moral or environmental goal; it's the only way to continued prosperity. "Our economic development remains limited by the high carbon structure of the energy sector," but "the challenges for the coal industry cannot be ignored, and, at present they are greater than the opportunities."[47]

Chinese leaders face an essential dilemma that is rarely made explicit: coal is irreplaceable for continued economic growth, but growth is limited by the pollution, inefficiency, and social costs of continuing to burn coal. The State Council, China's central governing authority, has decreed that coal use will fall. In itself, this is a huge step forward. Making it happen is another matter.

"There is a contradiction here," said Fan Baoying, the vice general manager of China Coal Technology and Engineering. "We must balance pollution control with economic growth. This is the next big issue for us."[48]

Since the Tiananmen Square uprising of 1989—which had its twenty-fifth anniversary in 2014—the PRC government has feared civil unrest above all else. Although China's state security forces have a well-earned reputation for quelling public protest, China is nevertheless one of the most restive countries in the world: the Chinese Academy of Social Sciences has estimated the number of street protests, impromptu or planned, at more than 90,000 a year.[49] Every day, someone in China is protesting against something: government corruption and environmental damage are two of the most popular causes. The Communist Party bureaucrats in Beijing have few illusions: they know that their legitimacy rests solely on the continued expansion of the middle class. If the rising masses can no longer afford to buy refrigerators, flat-screen TVs, automobiles, and air conditioners, the bargain that the party has struck with the people—growing wealth in exchange for political obedience—will no longer hold. China can't afford to shut down its coal industry quickly because China can't risk a recession.

What's more, the message—that the era of supergrowth in the coal industry has passed—has not quite gotten through to the big coal companies. In Shanghai I managed to corner Liu Jing, the general secretary of the Datong Coal Mine Group (also known as Tong Mei), one of the largest producers in Shanxi Province, China's coal heartland.

"We must recognize that the coal industry has limits," Liu admitted. "Globally the coal market is in decline. Our government is working to limit pollution and clean up the environment. This has forced us to make reforms and transform our operations. But we will continue to strengthen our facilities, and the amount of coal we produce will remain steady."

But demand and prices have fallen sharply, I replied. Doesn't that imply that Datong Coal must cut back, as well?

"No, no, the effects will be minor," he assured me. "We are one of the largest producers and we can withstand the market downturn. Before we were very focused on traditional coal mining, but now we are transforming and diversifying our business to include a combination of coal and electric power, coal-based chemicals production, and construction materials.

"We will keep all of our old enterprises steady while finding new areas of business. It's no problem, actually."

Onstage, Liu's boss, Zhang Youxi, the chairman of Datong Coal, was less sanguine. "The industry is under huge pressure," he declared. "Coal mining revenue was down last year by $11.5 billion, and the cash flow pressure is huge."[50]

China's response to all of these pressures is to migrate the coal industry inland. Datong Coal is consolidating its mines into 11 "clusters," or coal bases, each of which will have at least 10 million tons of annual production, 1,000 workers, and $1 billion in profit. Two of these bases, Ta Shan and Wangping/Xiaoyu, are already complete; together they produced more than $4 billion worth of coal in 2013. The coal bases will provide the cornerstones for "circular economy parks" that will include plants to convert coal into synthetic natural gas, or syngas, as well as cement plants, power stations, chemical facilities, and even "cultural tourism." The result, said Zhang, will be "the rebirth of the phoenix" of China's coal industry.[51]

This great migration—assembling mines, power plants, and industrial facilities into gigantic coal bases that will centralize the production and consumption of coal and send power to the east via huge ultra-high-voltage transmission lines—is "by any measure . . . the single largest fossil fuel development project in the world," according to *Inside Climate News*.[52]

"We will create multifunction power supply centers in the West, where power plants will use clean coal to provide power to the coastal cities," said Li Haofeng, the deputy director general of the Coal Industry Section of China's National Energy Administration, at the Shanghai coal conference.[53] China's relocation strategy has been evident for some time—it was first detailed by William Kelly in February 2014 in *Inside Climate News*—but Li's announcement was the first time it had been proclaimed officially to an audience that included Westerners.

To move all that electricity from the interior of the country to the coast, Beijing will invest nearly half a trillion dollars in ultra-high-voltage transmission lines built along west-to-east "power corridors." Earlier this year the world's largest UHV line started transmitting electrons from Hami, in Xinjiang Province, more than 1,500 miles east to Shanghai.

Relocating these industrial complexes to China's vast interior may make the skies over Beijing and Shanghai bluer, but it won't do anything about China's overall pollution problems. And in terms of carbon emissions and water constraints, it could make things much, much worse.

THE BALLROOM OF THE PUDONG SHANGHAI hotel was almost evenly split between Chinese attendees and foreigners. The non-Chinese coal executives from across the Pacific Basin tended to have world-weary expressions and an attitude of resigned exasperation toward the chaos they saw looming in China's coal industry.

"They're doing everything they can, pulling all the levers," Chris Atkinson, the Singapore-based market intelligence manager for the London-based mining giant Anglo American, told me. "But they are still trying to get lots of people out of poverty. People here enjoy the *thought* of green energy. But you can't just displace coal."

No industry is more dependent on what happens in China than the global mining sector. Shares in Anglo American, founded by Sir Ernest Oppenheimer in Johannesburg in 1917 to mine South African metals for the British war effort, have lost more than half their value since 2011. CEO Mark Cutifani has embarked on a radical reshaping of the company that includes selling off its South African platinum mines, which have been plagued by strikes, and shedding 20,000 workers. The plan does not, apparently, involve getting out of coal.

"You will hear that certain companies will avoid coal" going forward, Atkinson told me. "Anglo American is not one of them."

Those "certain companies" do include Melbourne-based BHP Billiton, the world's second-largest mining company, which has made a decisive strategic turn in favor of natural gas production, rather than coal. Even though it remains one of the largest exporters of coal, BHP Billiton executives have worked hard to separate the company publicly from the anemic market for thermal coal. In an indiscreet November 2012 interview with the *Australian Financial Review* that was seized upon by anti-coal activists, Marcus Randolph, at the time the head of the company's iron ore and coal division, described the future of coal as "very clouded." To the chagrin of his bosses, he went on: "In a carbon constrained world where energy coal is the biggest contributor to a carbon problem, how do you think this is going to evolve over a 30- to 40-year time horizon? You'd have to look at that and say on balance, I suspect, the usage of thermal coal is going to decline. And frankly it should."[54]

This fit of candor did not go over well at BHP's headquarters in Melbourne. Previously considered a contender for the job of CEO, Randolph was relieved of his membership on BHP's Group Management Committee and soon after went on an extended sick leave.

Still, Australia is China's biggest supplier of overseas coal. Second is Indonesia. Even closer to China's coal markets than Australia, the archipelago has seen a coal mining boom in the last decade. New mines have been opened not only by conglomerates launching multibillion-dollar projects, but also from small miners—often individual landowners with a backhoe, an open pit, and plenty of cheap labor. In the conference lounge at the Pudong Shanghai, I sat down with the Sharda brothers, Manish and Rishi. In their early thirties, the brothers run the coal trading company Virema Impex, headquartered in Jakarta, that was started by their father. Their suppliers are small miners scattered across the archipelago who load barges from island quays, bound for the coal ports of the South China Sea: Qinhuangdao, Tangshan, and Huanghua. This is Joseph Conrad territory: tramp freighters of dubious flag carrying unregulated cargoes of coal, and semiliterate mine owners doing business in fly-blown fish-smelling ports only one or two steps removed from the tiny satraps that for centuries ruled these provinces.

"We're working with professional [coal] buyers, and at the back with unprofessional miners," Manish told me. "Our job is to bridge the two."

"We don't know what happens at the back; we just know the vessel gets filled," added Rishi.

Often, the brothers said, they sign a 20-page contract with a miner who can't read. They buy in cash and sell on spec—meaning that if their buyers can't get a profitable price at the ultimate destination, Virema loses money. Virema sells coal free on board, or FOB, in Indonesian waters: the seller pays for the transportation of the coal to the port, and for loading onto the ship; the buyer pays the cost of shipping the coal, insurance, and unloading. The mining is entirely dig and scrape—"just making a hole," said Manish. They may sell 3 million tons of coal a year, a tiny fraction of China's coal addiction, but it's enough to create a profitable business. Or at least it was in the boom years.

"As long as the miners reduce their price, we'll be fine," said Manish, who was darkly handsome and incisive. "The question is, At what point is it no longer worth it, and they just decide to close the mine? For many of them it's just a matter of time."

As we spoke, news of China's coal bust was just reaching the distant islands of Indonesia. Many of the pits have opened recently; until the end of 2013 everyone with coal and a way to dig it was increasing their output. Reality set in abruptly. Indonesia's mining industry "is in panic mode," the chairman of Indonesia's Coal Mining Association told Reuters in July 2014.[55] Many of the small-time miners were shutting down, even as the big mining companies, such as Bumi Resources, Adaro, and Berau Coal Energy, boosted production in a desperate attempt to shore up revenues in the face of declining margins. As a result, total coal production in Indonesia was expected to stay flat in 2014, at about 463 million tons.[56] Indonesia's coal producers shared the religious conviction that permeates the industry: there will always be a market for coal. If you mine it, someone somewhere will burn it.

"Our suppliers are definitely behind the curve," said Manish. "When the market drops they get hit first. We as traders hope for the market to go up, but . . ."

"But we are realistic," said his brother. "There are always ups and downs in the market. It is survival of the fittest. Whoever survives will profit when the next market rebound occurs."

I'd heard the same words from Peabody CEO Greg Boyce, in his plush office overlooking the Mississippi River in St. Louis, three months before. *All we have to do is make it through this downturn.* The notion that this downturn might be different, that the coal industry had entered a slow, permanent decline, was unthinkable. Coal, the cheapest form of basic energy, will never be supplanted—or, at least, not soon enough to make a difference in our lifetime.

"The price will rebound," asserted Manish. "India and China will both continue to buy coal. The only thing that will limit coal, in the long run, is a technological advance in other forms of energy. In Asia, that is a long way off."

The great transition from coal to cleaner forms of energy seemed a long way off as the conference wound up and I boarded a night flight to Datong, in Shanxi Province. China seemed trapped by its own success. Reining in coal would require a level of sacrifice that the government, determined to maintain its tight but fragile grip on society, could not afford to demand. Like the proverbial man riding a tiger, China was afraid to dismount.

CHAPTER 7

SHANXI PROVINCE

On a Sunday morning outside Datong, at the northern tip of Shanxi Province, three old coal men took in the air outside their apartments. They roosted on a low wall outside a cluster of mustard-colored, five-story apartment buildings whose name translates, roughly, as "Harmony Village." The road that ran past the village continued up to the Qing Ci Yao coal mine. Dressed in Mao-era blue serge jackets and soft dark caps, they enjoyed a rare bout of sunshine beneath a sky washed blue by recent rains. In combined man-years they totaled about a century and a half underground.

I'd flown from Shanghai to Datong, China's coal capital, a grimy, vibrant city of 5 million people in the far north near the Great Wall and the border with Inner Mongolia. I was traveling with an entourage: a driver; a local interpreter who had adopted the English name "Michael" and spoke the Shanxi dialect, which is nearly incomprehensible to speakers of Mandarin; and a Shanghai-based guide from the venerable travel agency Abercrombie and Kent. We were on a weeklong coal safari through Shanxi Province, where the upheavals of the market and the government's anti-coal campaign are playing out on the ground and changing the lives of ordinary Chinese people. Driving west out of Datong, we'd passed a donkey car negotiating the clogged streets full of Toyotas and Hyundais and Volkswagens. For a couple of miles before the mine ran the walls of a large base of the People's Liberation Army. Outside the base gates a man was selling chow and mastiff puppies from a wire pen.

The retired miners were now beyond the reach of market forces. They had their pensions and they occupied small but comfortable apartments in Harmony Village. Their children and grandchildren lived not far away. They grew up nearby and went to work in the mine straight out of secondary school; like many Chinese mines, Qing Ci Yao is a self-contained community, with schools, housing, and shopping all within walking distance. Their long lives had been encompassed within the circle of the brown hills that were visible from the spot where they now sat.

White-haired and hale, with a kind, weathered face, Liang Sheng Cai said he'd spent "four or five decades" working underground. He estimated his age as 76, which meant he'd lived through the civil war, the Communist revolution, the Second World War, the Great Leap Forward, the Cultural Revolution, the unrest of the 1980s, and the long economic boom. So much history, most of which took place while he was underground, had made him philosophical. He seemed beyond anger or regret. "It was a hard job," he told me. "We worked deep in the mine, deep underground." There were three shifts: 7 a.m. to 3 p.m., 3 p.m. to 11 p.m., 11 p.m. to 7 a.m. The machines ran 24 hours a day. Today the mines are better constructed, more brightly lit, safer in every way.

Liang is healthy, even fit. Some of his fellow retirees have black lung, he said, some do not. He has watched the mines dry up or be closed since his retirement. "Ten years ago the provincial government started shutting down the small mines," said Liang. "The governor at the time had been a miner 10 or 15 years before. He ordered the integration of the small mines, and they kept the large, the state-owned mines. They wasted lots of coal, no longer running these mines."

He gestured up at the mine where he'd worked. "This is a local mine. The benefits are local, the income is reserved for this region."

While he spoke he fiddled with a length of string, wrapping and unwrapping and running it through his fingers. "The available coal is decreasing. It doesn't show up in the government production figures, but the coal is being mined out, here and at other mines."

In fact, China's official estimates of recoverable reserves are notoriously unreliable, and there are those who believe, and have produced studies showing, that at the current rate of burning the country will begin to run out of coal long before the government and the coal companies say, perhaps before 2030.

"The government is promoting other types of energy," he said, "solar and wind. The locals don't have the awareness of these forms. They just think: more output equals more money."

The friend at his side muttered something to Liang that Michael didn't interpret. Liang grinned. "We want to keep mining more to make more money!" he exclaimed, and his friends broke into laughter. "But I think the price will stay low for a long time. There's too much coal supply, much bigger than what we need."

It was a paradox. The reserves were being depleted for a glutted market. Liang has three sons and one daughter, all working for the mine in some capacity. "Only in the mines," he said. "Here there's no other choice."

"This mine doesn't work well anymore. The quality of the coal has gone down, and to mine it requires more technology, but they don't have the technology. Now they are late paying the workers. They just got paid for February. Not only they can't get paid, but the young people can't get jobs. The mines are fully staffed. There are no more jobs. What can the people do?"

Still, coal comes out of the mouth of Qing Ci Yao, and it gets loaded into trucks and transported somewhere. We drove on up the road in the warming morning and turned up the access road to the mine. A small gatehouse guarded the entrance to a coal yard where a front loader heaved piles of coal into dump trucks. In the time three huge haul trucks could be loaded at the mine face at North Antelope-Rochelle, a single dump truck was piled with coal and trundled down the hill. Two men, one wearing a white linen blazer that was only slightly soiled, greeted us. We had no appointment and no introductions. Here Michael for the first time told the story he would repeat a couple of times a day over the next week: I worked for an energy research company in America. (True.) We were thinking of investing in the energy industry in China. (Not exactly true, but not wholly false either.) I'd been sent to investigate possible properties for investment. (Not true at all.) Could we see the mine?

This required consideration. We stood in the yard watching the coal be loaded. "No photos," said the other man, who was wearing a company polo shirt. He waved at my iPhone and shook his head. Our unprepossessing appearance seemed to persuade them we were no threat, and after a while the blazered man waved us on, up the hill toward the mine works. A gray film of water, thick with coal dust, ran down the dirt road. Hills of coal 30 feet high stood along the road, and in a weedy lot opposite sat a pair of abandoned bulldozers that looked as if they dated from the Great Leap Forward in the late 1950s. A cluster of stone outbuildings crouched below the overhead conveyor that ran into the hillside beyond. We could not go into the mine itself but we could visit his office, said our guide. We accompanied him into a long two-story building with a series of dusty offices along a central corridor. In a cramped lab three women monitored machines that tested the qualities of the coal. The scene reminded me of old black-and-white photos of Marie Curie's lab. The man in the white blazer's name was Cao Guo Zhang, and he was the coal quality section chief for the Qing Ci Yao mine.

He ushered us into a cluttered office with a 2004-era computer and keyboard on a wooden desk that also held a couple of jars of coal samples labeled in Chinese characters. One of the female technicians poured scalding tea, unstrained, from a thermos. We sat and talked about the prospects for this mine, the coal industry in Shanxi, and the future of energy in China.

Just 28, Cao was the youngest section chief in his company, Datong Dimei. He was a graduate of the Chinese University of Mining Technology in Xushou, Jiangsu Province. His satisfaction with himself and his life's station was unmistakable. He barely stopped smiling during the half hour or so we visited in his office.

"This mine?" he said. "This mine could keep producing coal for another 50 years, maybe. We have stable output, our costs are decreasing, we have better equipment and better technology so we use less manpower. Or it could close next year," he said, grinning. "No one is sure."

Certainly Qing Ci Yao is one of the local mines that are next on the list for closing. It is small, outmoded, and inefficient. It's close to the city. As Liang pointed out, although it's part of a state-owned syndicate, it's not part of a larger mining cluster. Production had barely risen, from about 1.2 million tons in 1999 to 1.5 million in 2013, according to Cao. Automation has come slowly, if at all, to the small mines of Shanxi Province. Through Cao's window, on the hilltop above the mine, we could see a quartet of wind turbines turning slowly in the breeze like harbingers of a cleaner future. Cao's wife works at another mine, the great Ta Shan complex south of Datong. "I'm not worried about what happens with the mine," he said. A good capitalist, he had already hedged his future against the closure of Qing Ci Yao: he moonlighted as a moneylender. "I do business by myself, small banking, making loans to local citizens. Nothing big. Fair rates," he said, grinning.

Like many Chinese people, Cao has mixed feelings about coal's price fall. "In the short run it means my income is reduced. But it's better in the long run, better for the world.

"This mine is state-owned—the people here are not worried because they know the state will take care of them. That's the superiority of state-owned companies." It is also, of course, one of the primary defects of China's system: supported by the government, the state-owned companies are insulated from market forces. Surprisingly for a man whose livelihood depends on the production of coal, he supports imports: "We should use more foreign coal and less domestic, to protect our reserves."

Cao, though, is an optimist; he doesn't believe that coal could be in permanent decline. "The problems are temporary. The coal reserves are going down and so the price must rise. It's simple."

Having delivered this economics lesson, Cao took us on a tour of the mine's outbuildings. In the test lab the technicians recorded the ore's content: ash, caking, moisture, vapor, calories. Outside, a few workers in dark blue jumpsuits loitered outside a machine shop. In a warehouse-sized building, machinery coated in oily grime stretched to the ceiling. Made by Atlas Copco, the Stockholm-based supplier of industrial machinery, the equipment was only ten years old but looked as if it hadn't been used in some time. "Under maintenance," Cao kept saying. Light shafts fell from gaps in the high roof, and water had collected in puddles on the floor. The building had the profound stillness of places once filled with noise. I asked if the machines would ever run again.

"What will happen, who knows?" he shrugged, his smile unfaltering. "It depends on fate, and only God knows."

WE SAID GOODBYE TO CAO and drove south from Datong. The inner core of the city is ringed with high-rise condos under construction, tended by gigantic cranes. The former mayor of Datong, a man named Geng Yanbo, became nationally famous for his ambitious and costly program of road building, tree planting, and housing construction. Geng's plan was to clean up the city, replace the decrepit housing districts with modern apartment complexes, attract tourists, and reduce its dependence on coal. The condo towers were being built on a series of ring roads built in his tenure. Thousands of people were forcibly evicted from their homes to make room for his urbanization schemes. Geng, a rising Communist Party star, was demoted to vice-mayor in 2013, in an episode that was murky even by the opaque standards of Chinese politics. But his removal turned into a promotion when he was named acting mayor of Tai-yuan, Shanxi's largest city. The towers, and the cranes grouped around them like insect helpers, stood as a monument to Geng's tenure and to a quickly vanishing period in Chinese economic history.

"Who will live in all those new places?" I asked Michael.

"No one," he said. "No one can afford them."

Beyond the ring of empty towers, on Tong Quan Lu Road, we reentered the older China, passing through miles of ramshackle outskirts, past drab concrete apartment buildings, festooned with antennas, punctuated by stone hovels surrounded with rubble. A melon seller flicked flies from her cart, and stray dogs nosed among the crowds on the sidewalk.

The fate of the small mines might rest in the hands of God, as Cao had said, but for much of the province's 2,500-year history, He has evidently forsaken Shanxi. The borderland separating lowland agrarian peasant communities of the South and the warlike, pastoral nomads of the northern steppes, it was periodically swept by barbarian invaders and, in turn, the southern armies that repelled them. Two branches of the Great Wall, an inner and an outer perimeter, mark the northern boundary with Inner Mongolia, and Datong, the strategically important fortress city, for centuries supported numerous outposts along the frontier. One of Geng's programs was to restore the original city walls, which can be seen along the western edge of town. *Tong* in Chinese originally meant a hall or meeting place, but it has come to signify unity, harmony, similarity, or togetherness. *Datong* roughly translates into "Great Unity," and the major employer is Datong Coal Mine Group, or Tong Mei—"Harmony Coal." The capital has periodically changed hands depending on which group most recently invaded; as many as 300,000 imperial troops were stationed here at various times to guard the frontier. Shanxi was also the scene of one of the most appalling atrocities of the Boxer Rebellion, in June 1900, when 44 foreign missionaries and their children were lured by the provincial governor to Taiyuan and slaughtered.

Today, with some 35 million people, the province is one of the most polluted places on earth. Various government-initiated programs are underway to clean up the most affected areas, but the poisoned waters and choking skies will take decades to recover, if they ever do. (One paradoxical benefit of geography for the people of Shanxi is that the prevailing winds on the East Asian continent blow out of the north and west, and so carry the coal smog east to Beijing and the coast. There are days when the skies over Datong are blue and Beijing is smogged in—by pollution that originated in Shanxi.)

We came into the Ping Wang District, where Tong Mei has its headquarters. White-marbled, with high columns surrounding a grand entryway, it looks like a People's Congress building. The company is everywhere: we passed numerous outlets of the fast-food chain owned by the coal company: Tong Mei Food. The No. 1 Thermal Power Station, one of the numerous coal plants in the area, belched carbon dioxide–laden smoke into the sickly, yellowish sky. Above the road stretched sere hills, pocked by caves known as *yaodong*. Shanxi sits on the loess plateau, an arid region of dry, crumbly clay that has been called "the most erodible soil on earth." The hillsides are carved into fantastic shapes—pillars, domes, giants' fingers—and the hundreds of caves have been inhabited for centuries. Many still are. In the great earthquake of 1556, in neighboring Shaanxi Province, more than 800,000 people lost their lives when the *yaodong* they inhabited collapsed and entombed them. After the Long March, Mao and his Communist army sheltered in loess caves near the city of Yan'an, in Shaanxi, the birthplace of the revolution.

Rain is rare, and the poor soil supports only dryland crops like potatoes and corn. Coal has been burned here for centuries, but not until the industrialization of the mines and the production and transport facilities, begun under the Qing at the end of the nineteenth century, was Shanxi transformed into the country's energy storehouse. There are other industries in the province, but all of them, in one way or another, depend on coal.

At the city's far southern edge the inevitable coal loading station squatted on the north-south railway, surrounded by mini-mountains of coal. A conveyor painted the sky blue of Tong Mei slanted above the road. Beyond, the Yong Di Zhang mine, an outlier of the vaster Ta Shan complex to the south, sprawled across the hills. In the near ground was a temple complex dedicated to the warrior god Guan Yu, a deified general of the Eastern Han dynasty whose exploits

provide a central plotline of the *Romance of the Three Kingdoms*. Shrines to
Guan Yu can be found in shops and restaurants across the Chinese world, but
with his red face, his harsh and unyielding gaze, he seems especially well-suited
to Shanxi, a place stripped of illusion. After reversing course a couple of times
we stopped for directions to Ta Shan.

Mayor Geng is one of many politicians who have undertaken to turn
Shanxi into a tourist destination, and the province—once a cradle of Han
civilization—includes such famous sites as the Yunggang Grottoes, which
contain more than 50,000 Buddha statues from the fifth century, and the
picturesque Hanging Monastery. But no amount of face-lifting can disguise
Shanxi's essential industrial character: one-third of all of China's coal comes
from the province, and every shopkeeper and noodle-house owner in Datong
knows where their income comes from. There are more than 130 mining companies, from mule-and-shovel digging operations to enormous self-contained
coal cities, and already, by the time of my visit, the coal recession had claimed
some prominent victims. A former coal baron named Zhang Bao now runs
a potato company in Datong. In January 2014, an investment fund with the
confidence-inspiring name of Credit Equals Gold No. 1, formed to back a coal
mining venture called the Shanxi Zhenfu Energy Group, nearly defaulted
on its debt until it was rescued by an anonymous, almost certainly state-
connected, backer. This came a few months after the Shanxi coal tycoon Xing
Libin, whose lavish wedding party for his daughter became tabloid legend,
endured a spectacular fall when the coal company he founded, Liangsheng
Group, failed with 30 billion yuan, or $5 billion, in bad debts.

"The company's plight highlights a growing sense that China may be
facing the unraveling of distressed debts and assets, of which Liangsheng
Group may represent one high-profile example, after years of fast economic
growth and supercharged lending," reported the *New York Times*' "Sinosphere" blog.[1]

Coal's unraveling has fueled a nascent back-to-the-land movement in
Shanxi as coal miners return to the agricultural roots of their great-grandpar-
ents. According to the provincial Statistics Bureau, farm investments nearly
doubled from 2012 to 2013.[2] The government has supported this movement
with agro-friendly policies like price supports for certain crops and favorable farm loans. "The coal mine owners have become modern agriculture

trailblazers with their capital, technology and marketing skills," Guo Wei-dong, of the Shanxi Academy of Social Sciences, told *Xinhua*.[3]

Unfortunately, the limited amount of arable land on the arid plateau has been further reduced by rampant coal mining; many places in Shanxi are plagued by sinkholes and mudslides due to decades of uncontrolled mining. In early 2014 the province launched a six-year program to monitor sinkholes and landslides, at a cost of nearly $4 billion. It's hard to farm when the land is constantly giving way beneath you.[4]

It's also hard to mine when the price of coal is collapsing beneath you. Cao Guo Zhang's cheerful faith in the government's willingness to take care of the province's coal miners belies a paradoxical shift in the province's fortunes: even as privatization continues to engulf traditionally state-owned enterprises across China, the state is taking on a larger role in Shanxi's coal industry. The number of mine shafts in the province, once well above 2,000, will be cut to fewer than 800 by 2015, according to the state government. The remainder will be subsumed by the state-owned operators, their former owners compensated with company shares. Collectivization has gone out of fashion in modern China, but it's alive and well in Shanxi.[5]

As in capitalist economies, contraction has invariably led to corruption. The campaign to reduce illicit mining was accompanied by an antigraft campaign led by Jin Daoming, the former vice chairman of the provincial legislature and a Party stalwart. Appointed to head the of the Shanxi discipline commission, the main provincial anti-corruption body, Jin proceeded to use his position as cover for a campaign of personal enrichment. He was sacked and, as of late 2014, remained under investigation. His downfall "triggered an earthquake in Shanxi's halls of power," according to the business magazine *Caixin;* since the end of 2013 at least 18 senior officials in the province have been investigated for corruption.[6] Further clouding the Shanxi coal industry was the mysterious death, in 2011, of Xue Guojun, head of the finance department for Shanxi Shuguang Coking Group Company. The Shuguang group had acquired the Chuanwo coal mine and the associated Yuncheng Coking Plant under dubious circumstances, and Xue, fearful for his life, had reportedly attempted to resign more than once. His body was found atop the parking garage of an apartment complex in the southern Shanxi city of Yuncheng.[7] "Reports showed that Xue's body fell from a window above the fifth floor of a dormitory stairwell," reported the online news site *Caijing.* "There were deep marks on his neck and wrist and significant scars on his back."[8] In Shanxi, it's dangerous to work underground in the coal mines, and it's not so safe aboveground, in the office towers, either.

PAST THE CITY'S OUTER DISTRICTS, we came upon a region of brown fields with straight, tea-colored ditches bordered by rows of skeletal gray poplars. On the western horizon, under an ominous wall of clouds, stood the black-and-tan flanks of the Luliang Mountains; in the foreground were cemeteries with traditional mounded Chinese graves marked by simple stone markers. Interspersed with the graveyards grew small plots of scraggly looking castor beans and sugar beets. Muffled peasants toiled in the distance, using tools changed little since the fourteenth century. Everything was covered in a fine black grit. Woolen smog blanketed the sky, and the sun was just a rumor. We drove for a long time before the silos and conveyors of Ta Shan came into view.

The Ta Shan mine is part of the Ta Shan Circulated Economic Park, a conglomeration of industrial facilities about 20 miles south of Datong. The Economic Park is contained within the even bigger Jinbei coal cluster that stretches across much of northwestern Shanxi. Owned, inevitably, by the Tong Mei Group, the park comprises coal and iron mines; the Ta Shan Power Plant, which includes a pair of 600-megawatt coal-fired boilers; a methanol plant; a sewage treatment plant; chemical plants; its own railway; and a factory to make bricks from *gangue,* the nonflammable material that encases coal in ore. Under continuous construction since 2003, the Ta Shan park is touted as a new, environmentally friendlier model for the Chinese coal industry: excess heat from the power plant is used to heat employees' apartments, while coal ash is recycled as raw material for the cement plant. One plant's waste becomes input for a neighboring facility in "a garden-like new mining area," according to the Tong Mei website, that "beautifies the environment, clarifies the air, reduces pollution, and . . . creates a clear, comfortable and beautiful working environment for the employees."[9]

To a Western eye there is nothing garden-like about Ta Shan. There have been attempts at beautification: trees line some of the wider avenues, and the silos and conveyors were freshly painted in Tong Mei's colors: bright white and sky blue. But the place epitomizes China's toxic heavy industry: spread out below the mountains is a blasted plain of smokestacks, huge silos, polluted

streams, and savaged earth. Enormous towers for high-voltage electricity transmission march into the hazy distance like giant skeletal robots. Haul trucks
blast black smoke as they rumble down dirt roads. Ta Shan was the site of
one of the deadliest accidents in China's exceedingly deadly history of mining
disasters: in September 2008, not long before the new mine was completed,
262 people were killed when a dam holding back iron ore tailings burst and the
toxic sludge overwhelmed a village downstream.

Ta Shan also represents the long-term energy strategy that I'd heard described in Shanghai. In response to the country's environmental crisis, the central government is consolidating the far-flung coal industry, creating enormous
coal bases like Ta Shan, where, in theory, pollution can be contained, waste
recycled, and miners' lives safeguarded. The problem with this vision is that,
like Tong Mei's glowing description of Ta Shan, it doesn't capture the full
environmental havoc these coal clusters will wreak, both on the surrounding
areas and on Earth's climate.

By now it was midafternoon, and the day shift was getting off work in
the mines. We followed a truckload of miners, masked in black grime and
wearing blue Tong Mei helmets, piled into the back of a pickup for the ride
back to their base. Just off the road sat a cluster of low, narrow, yellow-brick
workshops, garages, and barracks. In the lot a labor gang toiled at the bottom
of a pit, replacing a sewage line. Climbing out of the truck, the off-duty miners
filed into a building with a group shower and a changing room where soiled
towels hung from banged-up metal locker doors. Wearily they shed their work
clothes, showered, dressed, and mounted aged scooters for the long ride to
their distant housing towers.

"For more than 100 years, generations of miners have dug coal for the
Qing Dynasty, warlords, Japanese invaders, the Kuomintang and the Communist Party of China, one after another," wrote Li Yang, a reporter for *China
Daily*, a couple of weeks after my visit.[10] There are roughly 200,000 miners at
Ta Shan, supporting another 800,000 or so family members. For generations
the mining families lived in small villages on the edge of the mines, or in
yaodongs in the loess hills. After the 2008 disaster raised an outcry over living
conditions in the area, Tong Mei constructed apartment buildings on the outer
parts of the industrial complex, cheap but serviceable, with company-supplied
water and power, basic medical facilities, and mostly paved roads. The coal

base is a self-contained world; everything needed to live, and to die, is contained within it. About 6 percent of the world's coal production comes from the land I could see from where I stood.

"They were born in the same hospital, went to the same school, eat in the same dining hall and speak the same dialect in the same accent," observed Li. "They're buried in the same graveyard."[11]

The *China Daily* reporter also maintained that the miners mistrust modern plumbing and that "their children share the same dream of working for Tong Mei as a miner, or marrying a miner." This, it seemed to me, confuses a dream with a default choice: until recently there were few other economic avenues open to the kids of Tong Mei miners. Miners' wages have traditionally been tied to the price of coal, not to volume or hours of work; when prices are high, an experienced miner could make up to 5,000 yuan a month, a comfortable living by Chinese standards. That's hard to match doing almost anything else in Shanxi. But that avenue, as Liang Sheng Cai had remarked, was no longer guaranteed. With consolidation has come mechanization, and mechanization means fewer, more highly skilled workers. As in Harlan County, Kentucky, jobs at Ta Shan are disappearing. The inexorable processes of modernization that have reduced coal industry jobs in the United States are slowly reaching Shanxi. Even if production keeps going up, there will be fewer miners making big salaries from Tong Mei.

As they dressed to leave, the miners regarded me and my posse with puzzlement. Outside, a small group of onlookers watched the men working in the pit, calling out an occasional suggestion. One guy was washing his VW sedan. Slightly apart from the others, a slender fellow wearing a dark blazer squatted on his haunches, smoking a cigarette and working two faux-jade balls through his fingers.

His name was Zuo Zhang and he had been working at Ta Shan for "five or six decades," he estimated. This sounded like a stretch. If he was over 60, he looked remarkably well preserved for a Chinese coal miner; he had handsome, wolfish features and black hair only partially streaked with gray. He spoke easily, pausing to form his words, and he had the cool reserve of a frontline military officer. I liked him immediately.

His job was to seal up mined-out areas to keep the toxic gases from leaking out; he headed a crew that ran bulldozers to plug abandoned mine shafts with concrete, stone, and riprap. Zuo hasn't retired because he's a supervisor and "they still need me," he said, chuckling. It sounded like a dangerous job, I remarked. He shrugged with a miner's disdain for risk and pulled on his cigarette. Like most Chinese coal workers I spoke with, Zuo was hardly ignorant of the global implications of what he did. "The climate is changing," he said. "We can see it here, every day." In the long run, closing small mines was a good thing. "In the future there will be no coal to mine. The land is getting hollowed out." Reducing pollution, and the carbon emitted from burning coal, was not just a desirable goal; it was an absolute necessity. "We can't keep going as we have," Zuo said, indicating his immediate surroundings. The consequences of doing nothing would be worse than those of transforming the industry, even if some of the men here at Ta Shan would be out of work.

"In the short run, it's troubling. I'm a laborer. Jobs like mine will be fewer. Some people will be forced to retire without pensions, to find other work." The men on the shift just getting off were mostly former itinerant farmers, forced from the land by the spread of coal and its associated industries. Their long-term prospects were uncertain. Zuo, on the other hand, had lifetime medical insurance and will be able to retire at three-quarters pay. It is the younger generation who will suffer in the new energy era. Zuo had two sons and a daughter. His youngest son was a student at Taiyuan Commercial College. His daughter was married to a tractor operator for Tong Mei. "It's good for him: he needs a job. Where else would he work?"

And his oldest son? "He lives with his wife and son in Taiyuan," said Zuo. He had a PhD in chemistry. He was a scientist at the Taiyuan University of Technology.

"What's his field?" I asked.

"Coal-to-liquids."

I'm on my way to Taiyuan, I said. Would it be possible for me to meet his son? Yes, yes, no problem. Zuo pulled out a mobile phone and, with Michael interpreting, read out the number. "Could we call him now?"

A few minutes later Michael was on with the son, Zuo Zhijun, who speaks some English. We made plans to meet the next day, when we arrived in Taiyuan.

"What do you think will happen?" I asked the elder Zuo. "Will China still be burning coal when your grandchildren are grown?"

"I think we can use coal more cleanly, more efficiently," he asserted. By now, this was a familiar response; almost everyone in the coal industry I spoke with said the same thing. "There is technology, in research and development, that could help reduce the pollution. But it's new technology. Not everyone wants to invest."

"And your son? Will his work help clean up China in the future?"

"I believe so."

"You must be proud of him."

He broke into a wide grin. "Yes, yes!"

On our way back to Datong we passed the No. 2 power station, an enormous generating complex on the southern edge of the downtown. Six smokestacks, hundreds of feet high, belched white, carbon-laden smoke into the sky to ride the winds east, and steam billowed from six cooling towers. I was 8,000 miles from my home in Boulder, but it occurred to me as we passed under the shadows of the plant that, ultimately, this place made electricity for me and my family. The steel in my Honda mini-SUV likely came from China; much of my clothing was manufactured, undoubtedly, in Guangdong; my iPhone was no doubt made in a Foxconn factory in southern China. The primary stumbling block for international climate negotiations has been the unwillingness of developing nations (i.e., China) to adhere to strict carbon emissions regulations imposed by the modern, consuming nations of the West, who have already passed through their industrial revolution and can afford to reduce their coal consumption. It's a false dichotomy. The coal mined at Ta Shan and burned at the power stations of Shanxi produces power for all of us. There's no escaping—we're all consumers. The only possible solutions are collective ones.

THE NEXT DAY WE LEFT DATONG and drove south on the expressway to Taiyuan. After a while I noticed that we'd left behind the sour, sulfurous odor of the city, and I realized that I'd seen very few surgical masks, the kind that are ubiquitous in Shanghai and Beijing, in Datong. The sere land stretched away, table-flat, marked by scrub. Camel-colored mountains stood off in the hazy distance, and vultures circled in the hot, cloudless sky. We turned off the highway and headed west, on the road to the mining city of An Tai Bao.

Passing car dealerships and small office parks, we came into the town of An Tai Bao, which, according to Michael, means "Appease Big Fortress"—mollify the army, you could say. Ascending into low hills, we came into the city itself, laid out in the valley to our left: blue and yellow apartment buildings, hundreds of pedestrians, shopping centers, schools, massive utilitarian office buildings. Tiny by Chinese standards and built from nothing to house the mineworkers, An Tai Bao has perhaps 50,000 people in a town originally designed for fewer than 20,000. We continued up the road to the security gate for the mine. We were, literally, at the end of the road, which ran up to and into the mine.

"Do you think we can get in?" I asked.

"Oh yes," said my Shanghai-based guide, who called himself "Jeffrey." "In China we have a more flexible way of solving the problem."

Michael and Jeffrey went into the low wooden office building that served as the welcome center at An Tai Bao. Security guards in uniform loitered by the open door, and knots of men smoked and waited and eyed our black SUV. In less than ten minutes Michael and Jeffrey came back.

"Twenty dollars," said Jeffrey. This was the going rate for a bribe into An Tai Bao. It seemed like a bargain to me. I handed over 140 yuan. A few minutes later we had a small placard to place on the dashboard and were driving past the raised metal gate. The dirt road, broad as a Beijing boulevard to accommodate the gigantic haul trucks, wound past big coal storage silos and then up into the mine proper. We came to another checkpoint and a well-fed security guard, straining the buttons on his tunic, came out to inquire about the purpose of our visit. Michael gave him our usual story: I was an "American coal specialist," considering investing in Chinese mines, etc. We had all the documents if he would care to see them?

He passed us through with an indolent wave. After another five minutes we came to yet another gate, which proved impassable because our vehicle lacked the flashing blue roof light and the triangular flag on a tall, flexible pole that would make us visible to the drivers of the haul trucks, whose tires were double the height of the SUV and whose beds could comfortably carry eight normal-sized cars. We continued on foot.

The open pits of An Tai Bao stretched out below the road and disappeared into the hazy distance. The sky above the mine glowed an infernal pinkish gray in the afternoon light. Lumps and boulders of coal, like the debris from a long-ago explosion, lay scattered at the side of the road, where we stood overlooking the operation. To our left was a crusher where coal was separated from raw ore, and before us, in a narrow valley, three conveyors carried the coal hundreds of meters to a conical processor five stories high, where it was sorted by the fineness of the rock. An endless stream of the black mineral rattled past, gleaming dully. I had the same sensation I'd had months ago at North Antelope-Rochelle, in Wyoming, as if I'd reached the dark secret at the heart of the matter, the brute central reality of our vast, industrialized, energy-hungry civilization, where men scrape from the earth the fuel to keep all of the boilers in all of the power stations and steel mills burning, night and day, continually. This was the place that fed the forges of the modern world, and that was slowly carbonizing Earth's atmosphere.

When it opened in 1986, An Tai Bao was the largest open-pit mine in the world. (It has since been supplanted by the merging of North Antelope and Rochelle.) An Tai Bao actually began as a joint venture between China's state coal ministry and an American company, Occidental Petroleum. Its financial origins are so densely tangled that it would become one of the famous case studies at Harvard Business School.

The mine complex was conceived in the 1970s, a time when the state-subsidized price of coal in China was one-quarter of the world free market price and fatality rates at Chinese mines were 50 times those at U.S. mines. More than one-third of China's coal came from small, unregulated mines that served local markets. The Ministry of Mines wished to modernize the industry, creating an up-to-date mining complex with access to transportation infrastructure in order to feed the burgeoning export markets in Japan, South Korea, and Southeast Asia. Here in the mineral-rich hills of western Shanxi Province, An Tai Bao would cover 17 square kilometers and produce 15 million metric tons of coal a year (it has since expanded many times over). The planners in Beijing knew that a project of this magnitude would require foreign money and foreign expertise. Both appeared in the person of Armand Hammer, the septuagenarian CEO of Occidental.

Hammer, the son of a Ukrainian immigrant whose deep involvement in the Communist Party in New York City earned him a prison sentence, spent much of his twenties in the Soviet Union, where he exported Russian goods to America and imported pharmaceuticals along with surplus U.S. wheat to help feed the millions of Russians who were starving in the series of famines that swept over the USSR in the decades between the wars. Returning to the States in 1930, he parlayed his early fortune into a series of oil investments that eventually led to control of Occidental Petroleum.

One of the few American businessmen to have done business successfully with Communist Russia, Hammer was determined to crack the vast China market, which opened to Western investment after Richard Nixon's historic visit in 1972. According to legend, Hammer showed up uninvited to a dinner for Chinese premier Deng Xiaoping on a visit to the States in 1979.

"We all know you," exclaimed Deng. "You're the man who helped Lenin when Russia needed help. Now you must come to China to help us."[12]

Within months, Hammer had visited China himself and signed deals for oil exploration and coal mining in the People's Republic. And Occidental emerged as the lead foreign partner in the massive An Tai Bao project. At first, China seemed to be the bonanza that Hammer envisioned. Occidental would put $400 million into the mine, all of which would be raised through "non-recourse project financing." In other words, wrote Carl Kester and Richard Melnick in the Harvard case study, "Occidental expected to own half the mine and be exposed to essentially no postcompletion risk."[13]

It didn't work out that way. Hammer, like many Western businessmen before and since, discovered that making deals in China was not as straightforward as it initially appeared. Protracted negotiations ensued, and Occidental was forced to put up $20 million in equity capital. Meanwhile, the price of coal fell from nearly $53 a metric ton in 1983 to $46 in 1986, when the deal closed. Hammer himself, wearing a full-length mink coat and a beaver hat, attended the groundbreaking for the mine, according to his biographer Steve Weinberg.[14] The expected revenues never materialized. Occidental extracted itself from the money-losing project in 1991, but An Tai Bao has been clanking along, producing coal ever since, with little regard for the ups and downs of the market, or for the environmental consequences.

In 2006 a *New York Times* reporter visited Shangma Huangtou, a village that had been engulfed by An Tai Bao, and described a hellish landscape wracked by earth tremors caused by excavation, rattled by coal trucks rumbling down the single road, its wells gone dry and its houses under perpetual threat of burial by the man-made mountain of coal waste that towered at the village edge. The villagers, the *Times* reported, had petitioned to abandon the town site. "We have no choice," said resident Wei Yangxian. "We have no water. The earth is sinking. The air is poisoned."[15]

One of three major mining operations owned by Pingshuo Coal Industry Corporation, China's largest coal exporter, the An Tai Bao mine, with its self-contained coal city, operates virtually independently of central government supervision or control. There's no law here except Pingshuo law. That presents the planners in Beijing with a paradox: Pingshuo is state-owned, but the state does not control it. Shutting down small mines like Qing Ci Yao is one thing; corralling the big producers like Shenhua, Tong Mei, and Pingshuo is another.

Huge operations like this make up at least 90 percent of China's coal industry, and there is little indication that they have been responsive to the central government's efforts to rationalize the industry. *Tian gao, huangdi yuan,* goes the proverb: "Heaven is high and the emperor is far away."

"There's a lot of new supply still coming on, mainly from consolidation, from the bigger mines with larger capacities," Ming Chang, a young British-educated trader whose family runs Fenwei, a coal consultancy based in Taiyuan, told me. "And none of them are willing to reduce their production. No one wants to be the first to take the bite. 'I'm waiting for the first idiot,' they say. And so far not many idiots have appeared."

These centrifugal forces have often determined the course of events. The tension between the center and the edges runs throughout Chinese history. Jonathan Spence, the dean of Western historians of China, wrote at length of these tensions, and how they played out specifically in the coal industry, in his masterwork, *The Search for Modern China:*

> In Shanxi province, where Deng Xiaoping had personally expressed an interest in using foreign technology to develop huge open-pit mines, the central government could not simply enforce its will over coal production as a whole . . .
>
> On many occasions, smaller mines might "hijack" rail cars for a week or two to move their own coal to local or national markets, only later returning them to the larger mines that technically owned the rail cars. A central decision to reallocate coal or open a major new mine was thus not a simple act . . .

The potential tensions between the center and the provinces, and within the hierarchy of each province, could have a paralyzing effect on state planning. Often the planning process itself had to run through a maze of channels before reaching the localities.

The rulers of the Qing dynasty had struggled for two centuries of their rule to streamline bureaucratic procedures, marshal and supervise errant bureaucrats, subordinate the provinces to the center, and defuse the social bitterness caused by corrupt behavior. . . . The PRC leadership, having tried to dissociate itself completely from . . . past abuses, now found that even the most advanced levels of technological planning were subject to the same tenacious tugs of localism and human frailty.[16]

Ultimately the planners in Beijing cannot have it both ways: either coal production and consumption must be limited, or the industry will continue running in defiance of government dictates and economic principles, like the conveyors at An Tai Bao, producing coal for which demand is shrinking. In Shanxi Province, coal had a momentum of its own, and Beijing was far away.

BACK ON THE EXPRESSWAY, we continued south to Taiyuan. A hundred kilometers south of Datong we began to climb through a series of tunnels that had been constructed with great loss of life, Michael told me. Scrubby pines dotted the canyon sides, and on the ridge above us stood a remnant leg of the Great Wall, rubbly and much reduced, unlike the restored sections near Beijing most often visited by tourists. There were no roadside towns, no truck stops, no fast-food chains with inviting signs. In Shanxi, outside of the big cities, where there is no coal there are no people.

Through a series of loess gullies we came into Taiyuan, marked by the arched bridges over the Fen River. One of the two longest tributaries of the Yellow River, it rises in the Guancen Mountains, in the north, and drains much of Shanxi Province. The Taiyuan Basin was once the scene of intensive cultivation watered by the Fen. In the 1960s, after a series of dams and reservoirs was constructed on the upper reaches, the river at Taiyuan went dry. Beginning in the 1990s the provincial government undertook an extensive restoration project, and today there is water in the riverbed again. A six-kilometer river park, with walkways, temples, gardens, and playgrounds, threads through the city center, and on the western shore a vast (and, when we visited, deserted) convention center, sports arena, and museum complex has been built along the river. No less aggressive than Datong, Taiyuan presents a more cultured face than its sister city to the north.

On busy Yingze Street, not far from the Shanxi Museum of Coal, we pulled through the gates of Taiyuan University of Technology. Founded in 1902 as the Western Learning School of the Grand Academy of Shanxi, one of the earliest modern universities in China, TUT has become the MIT of northern China, with 20,000 undergrads, most of whom are studying engineering or science. After a few phone calls we located Professor Zuo, whose father works at Ta Shan, and he came down to meet us on the sidewalk. After a quick foot tour of the thronged central campus we returned to his lab, a facility devoted almost entirely to the chemistry of coal.

"This is the Key Laboratory of Coal Science and Technology," he said as he ushered us past posters of the work that he and his graduate students are carrying out. Founded in 1984, funded by the provincial and the central government and by the coal industry, the lab has one central mission: to use applied chemistry to find a way forward to safely, cleanly, and profitably burn coal for another 100 years or more. This was not basic science. "We are focused on the targeted, pragmatic use of applied technology to supply technical support to the industry," said Zuo.

The lab's core concept is to raise the efficiency of conversion of coal into other forms that can be used in a variety of power generating and chemical processes. At just 32, Zuo, who graduated from a county high school near Ta Shan and won a scholarship to Taiyuan Tech, had discovered in his life's work a mission that is central to China's energy security: the effort to develop a process to convert solid coal to liquid fuel. Along with the amalgamation of smaller mines into huge coal bases, coal-to-liquids lies at the center of China's coal policy for the twenty-first century.

Simply put, China has plenty of coal but little petroleum (the country is trying to unlock its reserves of shale gas, but these are trapped in "tight," geologically challenging formations, and drilling technology in the country is at least a decade behind the United States). China is the world's second-largest importer of oil, behind the United States, buying more than 5 million barrels of oil a day. Much of that supply squeezes through the Strait of Malacca, the narrow passage off Singapore, between Malaysia and the Indonesian island of Sumatra, that forms the gateway between the South China Sea and the Indian Ocean. Chinese strategists are keenly aware that an attack, an act of terrorism, or an embargo that stoppered the Malacca jar would quickly squeeze the country's energy supply. Thus, building vast chemical complexes to convert coal to liquid fuels is not just an economic move; it's a geostrategic imperative. In January 2014 China's National Energy Administration announced a program to deliver 50 billion cubic meters of synthetic gasoline from coal by 2020—enough to supply one-eighth of the country's gas demand.

Unfortunately the price will be high. For one thing, it takes a lot of coal to make liquid fuel: about half a ton to produce one barrel. And converting the raw coal to liquid, which involves gasifying the coal by heating it without oxygen at superhigh temperatures, and then liquefying the gas via what's called the Fischer-Tropsch process, takes a lot of energy and a lot of water. A 2011 study by the National Resource Defense Council, drawing on earlier data from the RAND Corporation, found that the carbon emissions from gasifying coal, combined with the carbon emitted when the gas is actually burned for transportation or heating or electricity, doubles the total emissions that would be released from just using conventional gasoline.[17]

Meeting the National Energy Administration's coal-to-gas targets for 2020 would release an additional 12 billion tons of carbon dioxide into the atmosphere, and would far outstrip the water supplies available in the arid western provinces where the coal clusters are to be located. Many Chinese officials and scientists are confident that the majority of this smokestack carbon will be separated, captured, and stored permanently. But carbon capture and storage technology, which has been in the R&D phase for more than 20 years, remains uneconomical at scale. Economically viable carbon capture and storage, acknowledged Professor Zuo, is "many years away."

In the meantime, he and his lab are focused on a more attainable goal: creating a viable coal-to-liquids industry in Shanxi Province. Water, he admitted, is an issue: "The preconditions for any new coal project include water usage that is low enough to be sustainable. There must be an evaluation from the state water agency, or it won't go forward."

These were not theoretical issues for Zuo. He had a two-year-old son and his wife was an administrator at the university. He couldn't envision a future for China that doesn't involve coal, and he didn't see a sustainable future for coal using current technology. He knew that his father's decades underground allowed him to be where he was: a respected scientist at a prestigious institution. Once again, the story of coal was a story of fathers and sons. "As a student I didn't plan to be a coal scientist. I was very interested in chemistry, and one of my teachers was engaged in coal engineering. He led me into the study of coal."

I tried gingerly to ask him about the fundamental paradox at the heart of his work: trying to find a sustainable future for the least sustainable energy resource. "Do you feel good about the contribution your work could make?"

He paused and gave a long answer. "Every young person has their own hope or ambition. My goal is to one day work overseas in order to realize my dreams. I wish to be part of the worldwide scientific community, not just here in Shanxi. I'm not content with my current situation—I have higher goals."

I tried again. "Do you believe that your work will make a positive contribution, will make the world a better place when your son is grown?"

"Ahh. I understand. Absolutely I believe this. The coal industry, especially in China, is not so advanced in technology. I think we, here at this lab, are making a contribution to further the industry in many ways."

"In what way will your work contribute to making the use of coal less damaging to the environment?"

"If you look at the question, my individual research is not directly that helpful in reducing pollution. We are focused on developing the conversion steps, to liquid fuels, to gas, to synthetic products.

"But only in this way can we make coal cleaner and more sustainable. It will require many teams all contributing, step by step, to the ultimate goal."

"That's very admirable. I wish you luck. How close to that ultimate goal do you think we are?"

He chuckled softly, looking down. "I'm confident it will happen in my lifetime."

As if mocking his faith, the lights in Zuo's already shadowy office flickered and dimmed. I looked at the equipment on his cluttered desk: a couple of dusty desktop computers, circa 2000. If this was the center of the scientific quest for cleaner coal, it was hard to feel encouraged. Then, briefly, his ambition shone in the gloom. "You know, there have been two Chinese winners of the Nobel Prize in Literature: Gao Xinjian and, just two years ago, Mo Yan. No Chinese has ever won the Nobel Prize in the sciences."

He smiled modestly. This statement hung for a minute in the still office. The fluorescent ceiling lights flickered and revived.

EVEN BY THE STANDARDS of Chinese cities, Taiyuan has the frenetic energy of a perpetual boomtown. Despite its location on the invasion route of the Mongol raiders from the north, it gave rise to China's earliest banking system and to merchants who traded with both the nomads of the steppe and the lowlanders of the Yellow River and its tributaries.[18] It's a city of *arrivistes,* of unlettered coal tycoons, of sudden millionaires. Across the street from my hotel was a shopping mall with a Ferrari dealership and a Louis Vuitton boutique.

The day after we arrived I sat in a conference room of the Xie Tan Cheng Law Firm, in an office tower high above Taiyuan's gridlocked streets. Xie Tan Cheng literally means "Harmony Secures Peace and Success," but the English name of the company was more simple: "Talent Law Firm." At the invitation of the firm's founder, Wang Lifeng, I was taking part in what amounted to a seminar on the future of coal in Shanxi Province. The meeting had been arranged by Cao Xia, a professor of law at the Shanxi University of Finance & Economics who had been a student and protégé of Phillip Andrews-Speed, himself a professor at the National University of Singapore and one of the world's leading experts on China's energy economy. Around the table, fronted by printed name cards, were several lawyers from the Talent firm and a trio of experts including Gao Jianfeng, the former editor of the *Shanxi Economic Daily,* now a researcher at a Taiyuan think tank. We spent the morning talking about coal, coal-bed methane, environmental policy, and the harsh new realities of the dawning post-coal era. Listening to them, I sipped gingerly from the scalding hot glass of water—no tea, just water—that was constantly refilled in front of me.

"For 57 years, from 1949 to 2006, coal production in the province was consistent, around 500 million tons a year," said Wang. "After 2006 it increased to 1 billion tons a year. The coal companies have been slow to adjust. Most of the state-owned enterprises now have large debts. The Tong Mei

Group's profit is now essentially zero. The Jing Mei Group has 18 daughter companies, and their profits were zero last year. The policy makers have made mistakes. But the companies are stuck—they thought the boom would never end."

Wang was handsome, prosperous, and sleek. He wore an expensive European suit and an elaborate gold watch, and had a photo of himself at an Xterra off-road triathlon on the wall of his office. His daughter attended a Christian college in Tennessee. Founded in 2004, the Talent firm made its money in its early years advising the government on the reorganization of state-owned enterprises—the great wave of privatization that minted millionaires daily. More recently, the Talent firm worked for companies making energy investments: coal, oil and gas, and, especially, coal-bed methane, the gas nearly always found with coal deposits. Now, the province, like the country as a whole, was embarking on a great transition, from a coal-based economy to one fueled by cleaner forms of energy, less energy-intensive industries, more sustainable policies. But the change was too slow, Wang said, and it was too "administrative"—driven by government fiat rather than organic market shifts. Coal would remain the dominant source of energy until at least 2030, he said.

"The coal companies became so profitable that they became aimless. They didn't know how to invest all their extra money. Some raised pigs. Some sold air tickets. They made acquisitions everywhere. Corruption followed. Now that the price of coal has fallen they are selling off these enterprises. We need a stable economy, without so many ups and downs. We don't need so many high-rise buildings; we just want a peaceful life."

How exactly that peace would emerge was the subject of much speculation. The government would compensate laid-off coal miners; owners of coal companies could go into agriculture, like the former coal tycoon now selling potatoes; coal workers could install solar arrays. I thought about the dreams of Eric Mathis and Sustainable Williamson, in West Virginia. These schemes sounded equally fanciful. Consolidation would help; before 2006 there were more than 2,500 coal enterprises in Shanxi Province; now there were 1,053. I heard this oddly precise number repeated several times. Gao described the centralization strategy: the giant coal bases, the bigger, more efficient power plants, the extensive high-voltage transmission networks. I asked the same question I'd been asking since Shanghai: consolidating the coal bases would clean up Beijing and Shanghai, but what would it do to reduce overall coal emissions?

A pause, broken by uncomfortable sighs. "It's a major problem," said Liang Hu, one of the Talent lawyers. "Our first priority is development. In other words, to survive."

Survival is one thing; prosperity is another. Wang Lifeng is moving his firm into a commodity that many in Shanxi believe will provide the next boom: coal-bed methane. Also known as coal seam gas, coal-bed methane is natural gas found lining the inside of the pores in coal deposits. For centuries it was considered only a menace to coal miners. Boreholes, drilled from the surface, vented the methane, which has greenhouse properties far surpassing those of carbon dioxide. Only in the 1970s did CBM, as it is known, begin to be seen as a valuable energy resource. Today the United States and Australia both have CBM production industries, but it's seen as a less valuable, and harder to produce, alternative to natural gas found in shale deposits. In China, where the development of a shale gas industry has been much slower and more difficult than anticipated, and where coal deposits are still plentiful, it's seen as a vital national resource. Wang Lifeng and his firm helped develop the first regulations around the exploration and production of CBM, and he sees it as a source of revenue for his firm for years to come.

"In 2020 30 million vehicles will be gasified," said Zhu Ai Jun, the manager of a CBM refueling station on Yingze Street, not far from the Talent Law Firm offices. All of the taxis in Taiyuan now run on CBM, trucked in 300 kilometers from a coal mine. We were crowded into Zhu's office after the morning seminar and a ceremonial lunch with the Talent firm's principals. "The carbon emissions are 80 to 90 percent less than gas; there are fewer impurities, and the engines are quieter." A cubic meter of CBM has 13 percent more energy content than one of gas, Zhu said, and costs half as much.

The CBM vehicle program is a joint venture between the city and JAMG, a subsidiary of the Jing Ching Mining Group; the production of CBM at the mine is subsidized by the central government. Public buses are paid a subsidy of 3,000 yuan (about $489) to convert to CBM. Trucking the CBM to the city, in trucks that run on diesel, had to be a losing net-energy equation, even before taking into account the cost and energy required to convert the vehicles; but if CBM can be produced cheaply and transported via pipeline, it could become a viable alternative not only to gasoline and diesel, for transportation, but to power generation from coal.

The CBM boom, said Wang, would bring more demand for his firm's services: "The industry chain will be prolonged because of our efforts. Exploration, production, gas stations, end users . . . All will be our clients. We will make more profits!" he enthused. "And I will come visit you in America!"

This being China, though, there are complications. To produce CBM you pump fluids under high pressure into coal seams, not unlike hydraulic fracturing for shale gas, but the technology remains underdeveloped in China. What's more, ironically, coal mine owners tend to be opposed to CBM production even when it would bring them added revenue. "Many Chinese coal mine operators have opposed nearby coal-bed methane production, fearing that pumping sand and chemicals into wells to liberate gas might have the unintended effect of driving gas into their mines," reported the *New York Times,* a few months after my stay in Taiyuan.[19]

And bringing the CBM to big cities poses problems, as well: "New pipelines will be built on the edges of the cities," said Zhu, "but there are conflicts between the builders of the pipelines and the developers of condos. There are already many pipes underground, under these condos; if there was an explosion it would threaten many people. The cities' own development schemes contradict the development of CBM pipelines."

As a result, production has been slow to ramp up. In a July 2014 speech, the director of the National Energy Administration said that production of CBM would reach only 30 billion cubic meters in 2020, less than half the amount targeted just two years before and enough to supply less than 1 percent of China's electricity.[20] Wang's confident predictions aside, the CBM revolution in China is still decades away.

THAT NIGHT I WENT TO DINNER with Ming Chang, the consultant whom I'd met aboard the pleasure craft on the Huangpu. After my journey across Shanxi, that evening under Shanghai's softer skies seemed months before. Educated in Britain, Ming had repatriated only 18 months before. He held forth for more than an hour, in perfect, faintly British-accented English, on the future of coal in China as we sampled the distinctive Shanxi cuisine: flatbread, thick sliced noodles in aged vinegar, fried dumplings, lamb's head soup (which, thankfully, has nothing to do with the head of the lamb). Now in his late twenties, Ming had returned to Taiyuan to help run the family firm, Fenwei Energy Consulting, which, along with the global energy information company Platts, had developed the China Coal Index to track coal prices on the mainland. A merry, sardonic fellow, Ming had a view of the coal industry unencumbered by false optimism.

"It's worse now than it's been in ten years, but for us we think it's going to get worse still," he said. I asked how the turmoil in the coal industry reflected what's happening with the larger Chinese economy.

"If you gaze at the coal market, you automatically understand what is going on with China.

"For thermal coal, the end customer is power generation. And for coking coal, it's construction. Now construction has taken over as the largest consumer of electricity in China. If construction falls, then boom, there you go."

For Fenwei, though, there were still new business segments to be explored. "The coal market only opened up to foreign investment, and to global market forces, over the last 12 years. There are still new opportunities—for example, in financial derivatives. Particularly futures and swaps."

The Jungzhou commodity exchange started listing coal futures within the last 18 months, he told me, and the Dalian commodity exchange just in the last couple of years. "Coal futures are fairly new, and foreign capital is still not allowed to invest in them. We see these are real opportunities."

However beneficially those opportunities turned out for firms like Fenwei, the coming of speculative financial tools was not necessarily a good thing for the future of the coal industry in China. Peabody Energy sees a coming recovery for the world coal industry, I said. Is that likely in China? "Many people say that China is in a ten-year cycle of going up and down, and that the next upswing will happen later in this decade. But for that you need stimulation on a macro level. Now the government has become less willing to provide that stimulation. It's not sustainable: you see it getting so overbuilt, with all these luxury condos that nobody lives in."

Ming leaned back, wiped the noodle grease off his chin, and reflected on his country's tumultuous evolution. "China needed the short-term economic effects in the past, because people need hope. That's what the Chinese political system is good for, solving poverty in a short time frame. But that doesn't last forever. You can't sustain it."

In Shanxi, the government has made efforts to bring in new types of enterprises, in IT and services. So far those efforts have borne little fruit, for two reasons, Ming said. One is government incompetence, a time-honored tradition in China as elsewhere.

"The second reason is, there's no talent. Nobody with those skills in IT is going to move to Taiyuan. You want to swim in the pool where there's other swimmers of the same level.

"What is there in Shanxi beyond coal? Nothing."

CHAPTER 8

HANGZHOU

"It's not easy being Chinese," said Diwen Cao, sitting in the lobby cafe of the Ramada Plaza Hotel in the city of Hangzhou. She was referring to the place of China, as a nation, in the modern world, with its overpopulation, its typhoon-prone coastal cities, its environmental disasters, its overused resources and underdeveloped rule of law. But she could have been describing her own life experience, as a young, educated, and affluent Chinese person trying to find a way to make her country, and the planet, a better place. I'd met Diwen through the Hangzhou-based environmental group Green Zhejiang (named for the coastal province that includes Hangzhou), an organization she had parted with shortly before my arrival.

"At the moment I'm in between," she said. "I know what I'm good at. The good question is, What do I want to do? I'd like to find a position where I can help other people become change agents. . . . A lot of Chinese people are wondering, 'How can I be a more conscious person? How can I make a contribution?' I think I could help give them a better picture of the world at large, and an idea of how to make it a better place."

In her late twenties, Diwen, like Ming Chang, is a member of a generation unique in recent Chinese history: worldly, well traveled, socially conscious, and free of the xenophobia that still colors the worldview of many older Chinese. Young activists like her have led the environmental awakening that has washed over China in the last several years, challenging local officials and the central government, and rejecting the basic bargain that underlies the social contract in China: economic growth and personal prosperity in exchange for political obeisance and rampant environmental destruction. They represent the best chance for reducing China's dependence on coal and restraining the carbon emissions it produces. At the same time, they have the self-consciousness of people who live in between worlds, no longer completely Chinese and still seeking a suitable place in the wider world.

If there's a cradle for the birth of environmental consciousness in China, it's Hangzhou. In 2012 Hangzhou became the first city in China to declare that not only would it shut down the coal plants in proximity to the city, but it would wean itself off coal altogether. Hangzhou officials have pledged to source the city's entire energy demand from cleaner sources. The awakening of a collective environmental consciousness, said Diwen, could be heard in the dinner-table talk of her friends and their families. It could be seen in the numerous protests, some of which had turned violent, protesting environmental destruction across China in recent years. And it could be seen in the increasing unwillingness, on the part of local officials, to defy public opposition in order to expand coal's empire—in Shenzhen, for example, where officials, under pressure from the city's inhabitants, canceled the construction of a huge, 2,000-megawatt new coal-fired power plant in the city.

"The economic miracle has been based on the sacrifice of the environment," said Diwen. "If we don't change, whole areas, entire cities, are going to be unlivable. If people have alternatives, they will leave. But for the great majority, they have no choice. This is their home, they have nowhere else to go."

Diwen was among the fortunate who have alternatives. She was born in Hangzhou to civil-servant parents: her father worked for many years for the Chinese customs agency, the General Administration of Customs, and her mother was an accountant for the local government. Her father was a criminal investigator, tracking smugglers along the intricate waterways of Zhejiang Province's filigreed coastline. He came to know well the small coastal cities and the pollution-belching textile factories of the coast, and he saw firsthand the despoliation of the rivers and the bays. In those days, when Diwen was a child, no one questioned the worsening pollution, the leaden skies, the refuse-filled streams pouring into a sickening sea. And when AIDS became a medical emergency in China, few people were willing to look squarely at the origins of the disease and the factors in its spread. While still in university, Diwen cofounded a local campaign for the rights of AIDS patients and HIV-positive people. Humanitarian issues were her passion, and rather than going straight on to a master's program, as many ambitious Chinese grads do, she joined the Global Volunteer Network, becoming a nomadic aid worker in Nepal, India, and South America. She fetched up in Wellington, New Zealand, where she did her master's work in international development at the University of Waikato and became a protégé of Glyndwr Jones, a senior lecturer at the university's school of management. Jones, as it happened, was a Welshman who'd grown up in a coal mining village straight out of *How Green Was My Valley*. His influence helped convince Diwen that her calling lay in trying to save what was left of her homeland's natural environment.

"If you look at Taoist philosophy, the preservation of the natural world is one of the key concepts," she told me. "The Confucian values are all about balance, harmony, and sustainability. That's been the backbone of Chinese philosophy for thousands of years: to be in tune with nature. That's only changed in the last few decades."

She looked through the glass walls of the coffee shop at the crowds hurrying past. "In our quest to embrace the religion of money we've erased the essence of Chinese philosophy."

No city in China symbolizes that essence better than Hangzhou. Draped like a silk robe across the hills at the head of Hangzhou Bay, the city centers gracefully around West Lake, a magnet for poets, contemplatives, and sightseers since the early years of the Qin dynasty, around 220 BC. The completion of the Grand Canal, in 610 AD, helped make Hangzhou a prosperous center of commerce and culture, and during the period of Mongol emperors the city retained its status as a quasi-independent mini-capital, fed by the fertile croplands of interior Zhejiang. Both Marco Polo and Ibn Battuta, the great fourteenth-century Muslim traveler, praised it exuberantly, Polo calling it "greater than any in the world."[1]

The city was devastated (and nearly depopulated) during the Taiping Rebellion, in the 1850s and '60s, but recovered to be a primary stronghold for the Kuomintang during the civil war. Founded in 1897, Zhejiang University, Diwen's alma mater, is one of the People's Republic of China's elite schools and, like Harvard and MIT in Boston, attracts a vibrant population of students and academics who nourish the city's intellectual life and promote an atmosphere of free thought and dissent. Hangzhou has also been a haven of religious freedom, relatively speaking: the city has harbored small but vital populations of both Jews and Muslims in recent centuries, and two major mosques, the Great Mosque and the Phoenix Mosque, still stand within the city limits.

"As there is paradise in heaven, there are Suzhou and Hangzhou on earth," goes an ancient Chinese saying.

Even in paradise, though, rampant economic development and urbanization have swamped much of the city's former elegance. "Hangzhou doesn't have the benefit of being directly at the coast, with winds from the sea pulling away some of the smog, as is the case in Shanghai," wrote Andy Brandl, a German blogger and photographer who has lived in Hangzhou. "Rather, it is surrounded by mountains and has always been known for its foggy climate. When this natural fog inevitably combines with emissions, it is a recipe for a super-smog."[2]

China is home to 5 of the 11 supergiant, 5,000-megawatt-plus coal-fired power stations in the world, and three of them are located in Zhejiang—including the massive Jiaxing Power Plant, just up the coast from Hangzhou. Although the Zhejiang government has officially declared it will reduce and eventually eliminate the coal industry in the province, there are at least five new coal plants, totaling more than 8,500 megawatts, in planning or permitting, according to *SourceWatch*.[3]

Certainly, I never saw the sun during my days in Hangzhou. The enveloping gray lends a picturesque, soft-toned quality to the pagodas along the shores of West Lake, but that doesn't make it healthy.

And Hangzhou was the scene of one of the most notorious incidents in China's rich history of environmental disaster: the coal storm. On March 10, 2013, the famed city was inundated with black rain as a nightmarish downpour of wet, sticky coal dust covered every surface. "A thick layer of black ash covered pedestrians' hair, as well as the ground and the trees," reported the *Epoch Times*. Photos taken at Banshan National Forest Park, on the edge of the city, showed monuments and trees covered in coal dust. "A small white dog walking in the park also had black feet."[4]

No convincing explanation for the coal storm was ever made public. A few months later, man-made overdevelopment was matched by natural disaster as monsoon rains overwhelmed Hangzhou's inadequate sewer system and West Lake overflowed, causing the great flood of October 2013 that shut down the city for days and caused millions of dollars in water damage.

This combination—worsening pollution, a city in danger of ruining its famed natural beauty, and millions of independent, unruly residents—has led Hangzhou to confront directly the country's legacy of coal-burning and environmental catastrophe. In 2012, even before the coal rain and the flood, the municipal government announced the most ambitious coal elimination program in a Chinese city to date: by 2015, the plan stated, all coal-fired boilers in Hangzhou would be shut down or converted to natural gas or renewable energy. The "no-coal zone" would be extended, through 2017, to encompass the surrounding towns, and Hangzhou proper would get all of its power from cleaner and less-carbon-intensive sources of energy.

It was an ambitious goal, and it was unclear, from the official pronouncements, to what degree the city would continue to rely on "coal-by-wire"—electricity transmitted from distant coal-fired plants. But it was a bold attempt to envision a coal-free future for this once-gracious city of 2.5 million, the repository of so many of China's highest ideals.

Along with a handful of other cities with active residents and progressive-minded officials, Hangzhou has become the laboratory for an experiment in reinvention, an attempt to create a clean, modern, and sustainable city in the ruins of an older civilization: "We're trying to discover how to achieve that balance—to return to our roots and reclaim and preserve what our ancestors left for us," Diwen told me. "If it can be done anywhere in China it will happen here."

DIWEN CAO WAS VAGUE about her reasons for leaving Green Zhejiang. Her personal experience aside, the state of grassroots activist groups and cause-oriented nongovernmental organizations (NGOs) in China remains problematic. And Green Zhejiang was no exception.

Founded in 2000 by a young environmentalist named Xin Hao, Green Zhejiang existed for a decade in the unofficial limbo in which many citizen organizations operate in China. Xin, a native of Ningbo, had decided as a freshman at Zhejiang University to set out on an unlikely, not to say dangerous, adventure:

he spent 36 days bicycling some 2,000 kilometers to document environmental damage in the province. On his "2000 Eco Tour," Xin passed through 44 cities, photographing factories discharging untreated water into rivers, fish kills, and other damage. Seat-of-the-pants environmental activism like this was deeply discouraged in China at that time, but Xin Hao had a winning combination of naiveté and exuberance, and he was largely undeterred by official disapproval. The results were inconclusive—"We just reported what we found to the Environmental Protection Bureau," Xin told me. "I think some of the companies did clean up their wastewater." But the trip set him on course to create the most prominent and influential environmental group in Zhejiang Province.

Supported by a Ford International Scholarship, Xin spent three years at Clark University, in Worcester, Massachusetts, studying organizational management and making connections with U.S.-based environmental groups—particularly the Waterkeeper Alliance, which was eager to extend its activities in China. When he returned to China, Xin told me, "I thought I would just be a volunteer—I didn't want to make environmental protection my career. I wanted to make money!"

Xin is round of body and generous of spirit, and when he says such things he lights up with infectious laughter. I met him in Green Zhejiang's offices, a cluttered suite where desks sat nearly on top of one another; a banner with Chinese calligraphy adorned the wall next to detailed river maps, and a bottle of Johnnie Walker Black rested on a bookshelf cluttered with reports and books in Chinese and English. Originally called the Hangzhou Ecoculture Association, a rather innocuous sister organization to the Waterkeeper Alliance, the group finally won official recognition from the city in 2010. Registration at the provincial level followed in 2013, and the group adopted the name Green Zhejiang, as well as a broader charter and a more outspoken stance. "After 13 years we were finally legal," chuckled Xin.

Getting official recognition from the government has not come without tradeoffs. Diwen had hinted at some of them. This is true both for homegrown groups, like Green Zhejiang, and for the big international enviro NGOs, like Greenpeace and the Natural Resources Defense Council, as well. The Beijing-based groups know better than to involve themselves in local campaigns of street protest and civil disobedience; the grassroots groups know that they have a certain license to raise environmental concerns, with limits determined by local realities and government whim. Many environmental groups operate sub rosa, without license, a few people meeting in private homes for fear of being co-opted by official approval. Green Zhejiang, on the other hand, is largely government supported; under a low-carbon community initiative launched in 2013, the central government is providing select organizations with a total of more than 1 million yuan ($160,000) over two years to support government-approved activities—mostly public education, conferences, and feel-good ribbon cuttings on new solar arrays and the like. That's the major source of Green Zhejiang's funding. The rest comes from corporate donations, including the lease on the offices and staff salaries, plus a small amount, less than 100,000 yuan ($16,290) a year, in membership dues. In all, the central government gives domestic NGOs about 200 million yuan ($32 million) a year. That money serves as both a lifeline and a leash. But if Xin is troubled by the sources of his funding, he brushes it off easily.

"The central government is learning to be more open to NGOs and the will of the people," he said. "Years ago the people dreamed of being rich, and the government made that possible. But dreams change. Today the people dream of living in a good environment that will be healthy for their children. It's hard, right now, to imagine that—but before it was hard to imagine becoming rich, too."

In Xin's view China is moving from the period of "social management"—where, in effect, the government controls every aspect of people's lives—to "social administration," where the government works with corporations and NGOs to improve society, the environment, and individual lives.

"The government is trying and learning, and making mistakes," Xin assured me jovially. "In some cases they fail. But it's positive progress. They know they have to learn or face further violent protest."

True to Xin's accommodating nature, this is a generous view of developments in the country. The central government does not hesitate to crack down violently on protests that go beyond certain boundaries. This unspoken rule played itself out again in May 2014, just a month after my visit, when thousands gathered in Zhongtai, a resort town near Hangzhou, to object to the construction of a new refuse incinerator in their neighborhood.

Sporadic demonstrations had occurred for weeks at the site, where protesters had established a rough camp. Finally, on May 10, the scene erupted.

Chanting "We don't want the incineration plant! People of Hangzhou, save us! Chinese people, save us!," the protesters wrapped themselves in Chinese flags and carried banners that showed a skull wearing a gas mask, according to social media and news accounts.[5] Riot police moved in to disperse the protesters, first dragging off a group of elderly men and women who had lined up to protect the encampment. The protesters responded by tipping over police cars and setting them aflame, as the police attacked with batons, Tasers, and tear gas. Running street battles persisted through the evening. Official state media reported that 10 protesters and 29 police were injured, but posters on social media claimed that three protesters, including one child, had died, and scores were injured.[6] Although those numbers couldn't be verified, photos posted online showed unarmed demonstrators under attack and bloodied, with many taken to a nearby hospital.

The photos also show, better than any news article, the range and diversity of the protesters. Old women in quilted jackets, young hipsters with pop star T-shirts, children, middle-aged businessmen: the environmental movement in China, at least the not-in-my-backyard variety at work in the Zhongtai protests, has spread well beyond the young and educated, like Diwen Cao and Xin Hao.

The irony is that the Zhongtai plant, to be built by the Hangzhou Chengtou Group, was to form a key part of the government's coal-reduction scheme. The largest waste incinerator in Asia, it would burn municipal solid waste to generate power, helping to reduce the city's dependence on coal.

Already there are more than 100 such power plants, known as "waste-to-energy," or W2E, plants, in China, and the country plans to build up to 400 more by 2025.[7] W2E facilities could help solve two problems: coal dependence and the growing mountains of trash produced by China's burgeoning cities. The growth of waste has outpaced population growth in China's boom; in Hangzhou, the volume of solid waste is growing at about 17 percent a year, twice the national average, according to the *Financial Times*.[8] Modern W2E plants, in addition to generating electricity, are highly efficient and, relatively speaking, clean. W2E plants in China, however, tend to be cogenerators (i.e., they supplement coal boilers with burning trash), and official statistics on the amount of coal used and the power generated from refuse are likely to be inflated. Nevertheless, W2E could provide a source of effectively renewable energy with less carbon dioxide emissions than coal-fired plants. But that's hard to explain to a group of citizens who don't want a huge refuse incinerator built in their neighborhood.

In the wake of the riot at Zhongtai, officials promised to review the plans for the plant. Such promises, in China, are often hollow, and there was little indication that construction would be halted. In other incidents, though, public pressure has managed to halt the construction of new coal-fired power plants. The most famous, and significant, case happened in the traditional fishing town of Hainan.

SEPARATED FROM THE MAINLAND by the narrow Qiongzhou Strait, the island of Hainan hangs off the coast of southern China like a pearl from an earlobe. Much closer to Hanoi than Beijing, the island, for centuries a part of Guangdong Province, has often been treated by the state as a foreign colony; in imperial times it was used as a place of exile for disgraced officials, and the island has always been home to a rich stew of ethnic peoples who speak languages incomprehensible to mainlanders: the Li, the Miao, the Limgao, and, at Sanya Bay, a small population of the boat-dwelling Tanka.

Coming to Hainan from the mainland, it's easy to think you've arrived in a Southeast Asian island country—Indonesia, perhaps, or the Philippines. China's only tropical province, Hainan is blanketed by oceanic fog in the winter. In the central highlands, volcanic peaks rise above dense tropical rainforest, and the island is ringed by largely undiscovered beaches. Although 9 million people inhabit the island, and Haikou, the capital, has more than 2 million people, most of the towns on Hainan are, by Chinese standards, little more than villages.

Until recent years the major industry on the island was agriculture; the Hainanese grow rice, coconuts, sisal, and pineapple, and there are large rubber plantations in the central highlands. More recently an ambitious government campaign has drawn large numbers of tourists to "the Hawaii of China." Since 1988 Hainan has been a special economic zone—a privileged economic enclave where free trade is allowed and encouraged. A flood of real estate investment and shoddy development schemes ensued, leading to a massive property bubble that crashed in 1990. Growth since then has been steady. The island now has nearly three dozen five-star hotels and a new convention center that cost nearly a quarter of a billion dollars.[9] The annual Bo'ao Forum, the "Davos of Asia," draws hundreds of plutocrats and politicians to Bo'ao port on the island's east coast. But even today, what Hainan is known for is fishing.

The waters of the Gulf of Tonkin, to the west, and the South China Sea, to the east, were for centuries so choked with fish as to become mythical: it was said that you could walk from Hainan to the Vietnam coast, stepping on the backs of fish. Until overfishing decimated their populations, seemingly infinite schools of bonito, mackerel, tilapia, and tuna, as well as shrimp and crab along the shores and shallow bays, provided the Hainanese with good livelihoods and a healthy export business. At the island's southern tip, the "End of the World" to mainland Chinese, stands a statue of a fisherman gazing out at the South China Sea.

Like inhabitants of isolated fishing communities the world over, the residents of Hainan are prideful, self-reliant, resistant to change, and very stubborn. Armed resistance—to Han Chinese intruders, to the Japanese in World War II, to the Kuomintang during the civil war—has a long and celebrated history on the island. And in 2012, when the government announced a plan to build a massive coal-fired power plant at Yinggehai, on the island's southwest coast, old instinct kicked in. Pollution and industrial fishing had nearly destroyed the traditional way of life in the coastal villages: the island's forest cover had been reduced by more than 90 percent, and the mangrove forests that sheltered the shore from typhoons were rapidly disappearing. The island was becoming a playground for wealthy mainlanders and foreigners. The Hainanese had had enough.

On April 11, 2012, a demonstration against the plant at Yinggehai—a former fishing village populated by the Li people—drew close to 10,000 people who feared the remnants of their fishery would be wiped out by pollution from the plant. Shocked by the fervor of the anti-coal protesters, officials retreated. Eventually they decided to relocate the project elsewhere on the island. But residents of two other proposed towns revolted as well, and after several months the project reverted to Yinggehai. This time the protests turned violent.

A ribbon-cutting ceremony planned for October 16 was disrupted by thousands of protesters—"half the town," according to one account—who prevented the event from proceeding.[10] "Hundreds of the town's women subsequently gathered in front of the local government building," reported the online news site *WantChinaTimes*. "Others surrounded the entrance of the local fishery bureau day and night to prevent the authorities from taking out the plaque and finishing the ribbon-cutting ceremony."[11]

This time the authorities were prepared: wielding truncheons and firing tear gas, the police dispersed the protesters and arrested hundreds. Police barricaded the town. Sporadic protests and rioting continued, though, and the authorities again declared a pause to the project.

Like coal supporters everywhere, the provincial government and Guodian Ledong Power Generation Company, the state-owned utility funding the project, pointed to an ineluctable fact: Hainan needed more electricity. With tourism expected to expand nearly 13 percent a year through 2021, and the island's existing hydropower capacity already outstripped by demand, the plant was needed to support economic growth on the island.[12] The potential for renewables on the island was low. Once again, the only answer was coal.

While local officials tried to appear willing to negotiate, their actions spoke louder: Hainan environmentalist Liu Futang, who had given a face and a voice to the Yinggehai protesters and whose writing had long documented the environmental damage of big government projects, was arrested and charged with "illegal business activities."[13] Even as the protests continued in Yinggehai, Liu was convicted, fined $2,800, and sentenced to three years of probation.

"Hainan is a real-life example of that film 'Avatar,'" Liu told the *Washington Post* in 2010, speaking of the rampant development brought from the mainland. "Except in 'Avatar,' they could organize together to fight back." Of Hainan's future, Liu remarked, "I don't have much hope—nothing can stop this change."[14]

That Liu avoided jail time could be seen as an act of relative clemency from a government that has dealt harshly with environmental activists in recent years. But his conviction sent a clear message. Further street protests would not be tolerated. The central government's National Development and Reform Commission was petitioned by Hainan activists to cancel the plan. In response, the government pledged to conduct a thorough environmental review of the project. But the coal plant had too much momentum, too much money behind it, and too much support from the state to be derailed. In opposing the Yinggehai plant the people had discovered their voice. But the government proved once again that it had the will, and the force, and the staying power to outlast popular protest.

THE OUTCOME WAS DIFFERENT in Shenzhen, 300 miles to the northeast in booming Guangdong Province. There, a year after the anti-coal protests erupted in Yinggehai, a coalition of environmental groups, business leaders, and local officials came together to actually kill a planned coal plant.

Nowhere in China is the air more polluted than Guangdong, which in the last 20 years has become the factory and warehouse not only for China, but also for much of the world. On my first day in China I took the fast train from Hong Kong to Shenzhen, north up the Pearl River Delta, through wetlands dotted with paddies, fish ponds, clusters of traditional houses, and, every 20 miles or so, a gigantic coal-fired power plant. In the haze stood the factories that had fueled the Chinese miracle. Open steel-beam towers carried ultra-high-voltage transmission lines in every direction, crisscrossing under the gray pall of smog. The air smelled metallic. Visibility was half a mile. Each time the train crossed another broad, sluggish tributary of the Pearl, the far shore was invisible. Along with Ta Shan, it was the most heavily industrialized and polluted landscape I'd ever seen. Guangdong is where we have shipped the dirty work of modern society.

A fishing town of 30,000 only 30 years ago, Shenzhen now has 10 million people and a GDP of $200 billion—roughly equal to Peru's. No one came to live in Shenzhen because they wanted to live in Shenzhen; for that they went to Shanghai, or, if they could manage it, Hong Kong. Still, a rough frontier ethos had grown up in the sprawling megacity—including an expat community that tended to live and play in the Shekou district and liked to make rueful jokes about each day spent living in Shenzhen erasing a month from their expected life span.

Shenzhen Energy Group had been planning for nearly eight years to add a major coal-fired plant in the city, but when the plans were announced, in early 2013, the response was immediate and overwhelmingly negative. Though it was to be equipped with the latest scrubbers and filters to reduce mercury, sulfur, and nitrous oxide, the 2,000-megawatt plant would be within 50 kilometers of more than 17 million people and, according to Greenpeace, would cause

1,700 premature deaths over its operating life and would, moreover, further clog the skies over South China with coal smog.

In this case the resistance to the plant was expressed not in street riots but in a campaign launched on social media by local activists and intellectuals—one that spread rapidly to the popular press. After that events moved quickly: 43 deputies of the Shenzhen People's Committee signed a petition that called on the municipal government to reconsider the plan. It did not help that the plant had been officially approved by Liu Tienan, a National Development and Reform Commission official who'd since been placed under investigation for corruption and keeping a mistress. Within weeks the municipality formally asked Shenzhen Energy Group to cease preparation for the plant. What's more, the city declared it would prohibit the construction of new coal plants within the city limits. This swift resolution happened despite a call from the central propaganda bureau, in May, to suppress coverage of the plant: "Firmly reign in the degree of coverage of the Shenzhen Binhai power plant," the bureau, often referred to by its Orwellian nickname, the "Ministry of Truth," commanded Chinese media, according to a report in *China Digital Times*. "Prevent malicious sensationalization."[15]

Sensationalized or not, the campaign appeared to be a rare and unmitigated victory for the forces of modernity and conservation: "This is the first project that has been cancelled mainly on the basis of concerns about air pollution," wrote Greenpeace's Justin Guay. The successful shutdown, he exulted, has "bent the arc of history in China."[16]

Indeed, it seemed to be a remarkable acknowledgment of the need for more transparency and public input on large infrastructure projects. "For a local government, it is not a matter of just dissuading local residents from opposing the project, it is a matter of establishing their own trustworthiness among local residents," declared the *China Daily,* which often reflects the official line out of Beijing. "Many local governments are yet to develop the habit of communicating openly with residents. Neither do most of them have the awareness that transparency can be a way to reduce and dispel the mistrust between them and residents."[17]

In truth the Shenzhen victory was more alloyed than that. The plant was not canceled, as many Western outlets reported; it was merely relocated. Shenzhen Energy said it would build the plant elsewhere, farther from the eyes of the public and the press. As with the plan to build coal bases in the interior, shoving coal plants under a rug of official secrecy is not going to solve China's overall pollution crisis, much less help reduce global carbon emissions. Still, the movement in Shenzhen had demonstrated an important truth: public will could, sometimes, divert the implacable forces of state-supported coal.

"Guangdong still has a large pipeline of new coal-fired power plant projects, but the public concern on air pollution and pressure to stem coal use will make these increasingly unlikely to be implemented," declared Lauri Myllyvirta, a Greenpeace energy campaigner, optimistically.[18]

WHEN ZHAO ZHONG CAME TO Lanzhou in 2004, as a freshly graduated nuclear engineer with the Institute of Modern Physics at the Chinese Academy of Sciences, he had little concept of the place he would now call home. Zhao had grown up comfortably in Hefei, in Anhui Province. At Hefei University of Technology, he joined the first generation of young Chinese people to become active environmentalists; an avid outdoorsman, he began to see that "the mountains were dying," he told me, and he helped launch Green Anhui, one of the new crop of grassroots environmental organizations that, under Premier Zhu Rongji and his successor, Wen Jiabao, enjoyed a period of new freedom of action in the late 1990s and early 2000s.

At 22, Zhao was struck by the harsh beauty of the landscape surrounding Lanzhou, in western Gansu Province, where the Gobi Desert, the Loess Plateau, and the Tibetan Plateau collide. The Yellow River, which flows through Lanzhou, waters the semi-arid province sparingly, and the bare Qilian Mountains, reaching to more than 15,000 feet, scrape the cloudless sky. For centuries Gansu provided a primary corridor for the caravans of the Silk Road, carrying Chinese silk, pottery, and jade west and gold and silver to the east.

But Lanzhou was one of the most polluted places on earth. Petrochemical refineries and mining had released heavy metals into the water, which poisoned vast tracts of farmland. Coal was mined in Gansu, but the province's primary output was rare earths, used in many high-tech applications including missile guidance systems, mobile phones, and batteries for electric vehicles. At one point tests found that the Yellow River at Lanzhou was 10 percent sewage. Breathing Lanzhou's air, it was said, was like smoking a pack of cigarettes a day.

Zhao Zhong had to do something. He created Gansu's first environmental organization: Green Camel Bell, named for the tinkling decorations worn by camels trodding the Silk Road. With a small grant from San Francisco–based Global Greengrants Fund, he set up an office and assembled a skeleton staff, mostly student volunteers. They began by doing what small local NGOs do in China: organizing trash cleanup drives, handing out flyers, putting on workshops and teach-ins. Raising awareness seemed enough: the concepts of conservation and preserving the natural environment were new to Gansu. "When we visited communities," Zhao recalled in a 2012 interview, "they would ask: 'Why are you here?'"[19]

But Zhao soon realized that public education alone would not suffice, and that he had a unique set of skills and a unique opportunity: recently passed information disclosure laws required polluters to disclose the substances they were pumping into the atmosphere and the water. Those laws were largely ignored, but Zhao, a skilled technician, began using a GPS system borrowed from a local university to pinpoint factories that were dumping waste into the Yellow River. This was a new tactic in China, and, surprisingly, the provincial government decided to enforce the law. Spotlighted by Green Camel Bell's mapping efforts, the polluters were forced to clean up. Green Camel Bell's maps became the basis for China's first national Water Pollution Map. It was a signal victory in China's nascent environmental movement, and it made Zhao Zhong famous.

In 2009 *Time* named Zhao a "Hero of the Environment."[20] In China, he became an adviser to local environmental groups across the country, and he was in-demand as a speaker to U.S. audiences as well—an activist allowed by the state, for some reason, to pursue his work, an enviro-warrior who'd actually scored significant victories in the battle to save China's land, air, and water.

Now the China program coordinator with Pacific Environment, based in Beijing, Zhao has taken on a larger and in some ways more fearsome dragon: coal. "The coal program is still very new to Pacific Environment, we just started this work in 2012," he told me in early 2014. "The coal and energy issues are very serious and important, and we hope to use the same strategies that have been successful in combating water pollution."

By targeting specific coal-fired power plants, Zhao said, he was trying to work with the government, to improve the energy policies at the national and provincial levels and to tighten controls on smokestack emissions. He was also embarking on his own coal safari: a series of visits to coal bases in the west and the north to collect firsthand evidence of the damage from these vast but remote facilities, which most Chinese people in Beijing and Shanghai will never see and know little about.

"Through this trip we want to show our partners that there is reason for hope that we can improve the coal campaign. We need firsthand evidence of what's happening in Shanxi, in Inner Mongolia, to help people understand why it's very urgent and important to control these developments before they are fully built and out of control."

The problem, Zhao said, is the tangled local connections that link coal miners to power producers to local officials in a mutually beneficial but ultimately disastrous web of business, power, and influence. In Gansu, the local and provincial officials had sided with Green Camel Bell against water polluters. In coal country it is different. "If you hope to fight against the coal power plant, in other words you are fighting the local government. I don't know how to avoid this. It's very sensitive, and we hope to develop some strategies that will help us be effective." What would success look like? I asked him.

"I don't see that we will stop the coal mines, but we will promote people's health, and document the damage to air and water, and take care of reducing the number of people who are impacted by air pollution and coal ash. That I think is the best we can hope for, for now."

Is Zhao, now a veteran of China's environmental struggle, optimistic about the future? "I believe the central government is truly serious about reducing coal. The leaders are in Beijing, they are also victims of air pollution. With water it's easy for them to have purification equipment, but when the air is unhealthy everybody is the victim.

"But we must move step by step. These are still very sensitive issues. I can say from my experience with water pollution that the first thing is to make sure that the people understand, through education and awareness, that they should have a right to a good environment. So that when they learn about some coal power plant that is being built in their area, that the nearby communities will band together to protect their rights. We have seen this happening already, in Hainan, in Shenzhen."

That's the first step, Zhao said. The second step will come when the political environment improves—"when many citizens have these moments of environmental activism, they become involved in these movements and they try to organize these kinds of activities." The water pollution campaign, which has seen some notable successes over the last decade, could provide a template for resistance to coal. "Eight or ten years before, water pollution was also a very sensitive issue, like the coal campaign today. So many people spoke out about their local rivers and their water that the government was forced to acknowledge the problem, and to take steps to clean up the rivers. We can foresee that, maybe in four years or even sooner, the coal campaign will likewise be supported by the public and by the government, that more and more grassroots NGOs will spring up against coal and that many local activities will force the government to take action.

"That's my hope."

"AN ANT ON THE MOVE accomplishes more than a dozing ox," said Lao Tzu. It's tempting to see the nascent environmental movement in China as an army of mobile ants, and the coal industry as a sleepy ox. And it's true that there are plenty of indications that the ants are making headway.

With nearly 1.2 billion yuan ($200 billion) invested in wind, solar, and biofuels from 2012 through 2014, China is by far the world's largest investor in clean energy.[21] The amount of energy China obtains from nonfossil sources has more than doubled since 2005, and the economy's energy intensity, following the Kuznets Curve, is falling steadily. As Greenpeace's Myllyvirta observed, local victories like the one in Shenzhen are multiplying as the people find their voice and the central government grows increasingly reluctant to foist dirty energy projects on unwilling subjects.

"The first step toward sustainable economic growth is to recognize, as China's leaders have, that pollution—produced largely through coal-fired power plants—is profoundly damaging citizens' lives and livelihoods," wrote

He Jiankun, director of the Energy, Environment, and Economy Research Institute at Tsinghua University, and Nicholas Stern, a member of the British House of Lords and a professor at the London School of Economics, in a July 2014 essay on the Project Syndicate website.[22] It's safe to say that this recognition would not have happened, or that it would have come much more slowly, without the courage and fervor of environmental activists like Liu Futang and Zhao Zhong.

In September 2014 at a conference in Tianjin, China's National Development and Reform Commission announced a plan for a nationwide carbon market that could regulate up to 4 billion metric tons of carbon dioxide a year and be worth up to 400 billion yuan ($72 billion) by 2020, according to officials. Along with President Obama's plan to regulate carbon dioxide emissions from power plants, announced three months earlier, such a market could represent a turning point in the struggle to stop the growth of greenhouse gas emissions.[23]

One way or another, coal demand in China has almost certainly peaked as a percentage of overall energy use, though it is likely to keep growing in absolute terms. Despite the recalcitrance of the big coal producers, the restructuring of China's coal industry is surely inevitable, as is the slowdown of the massive proposed coal mining projects in Australia, Indonesia, and the United States to feed China's coal addiction. As always in this vast, vibrant, and endlessly contradictory nation, contrary evidence abounds—most especially the establishment and expansion of coal bases, like Ta Shan, in the interior. But peak coal in China is happening, almost before our eyes. The question is not whether the government can reduce China's reliance on coal; that is going to happen due to a combination of public opposition, market forces, and official policy. The questions are whether this can be accomplished with a minimum of social unrest and in enough time to have a noticeable effect on global carbon emissions.

"Gaze at the coal industry and you will understand China," Ming Chang had told me in Taiyuan. I thought of that remark on my last day in Hangzhou. I had an appointment in an office tower next to a riverside park on the bank of the Qiantang River, above its confluence with Hangzhou Bay. Afterward I walked down to the river. The sky, of course, was gray, but the day was mild, and for once the acrid smell of coal was absent. A conference of young people, perhaps a career fair, was just breaking up, and the attendees drifted from the tents to the grassy riverbank. A pedestrian bridge crossed the slow-moving river; on the other side was an imposing skyscraper that resembled a futuristic, vaguely menacing pagoda.

I stood on the bridge and watched as a barge, 30 meters long and powered by a small motor launch, rounded the upstream curve. In low mounds on the deck the cargo was piled under plastic tarpaulins. I wanted to think it was coal, being barged to the coast as it had been for hundreds of years. More likely it was gravel. The boat engine chugged *basso profondo* as the barge swung past and headed downstream, disappearing into the smog.

When I returned to the States, people asked me whether China is serious about getting off coal and cleaning up its environment. The answer, unquestionably, is yes. Then they asked if my trip had made me more optimistic, or less, about our chances of shrinking the coal industry and slowing climate change. That's a more complicated answer.

What I saw in China was a society in the throes of rapid and unprecedented change, resisted by a government whose rule is based in stasis. I saw a dawning environmental consciousness—a return, as Diwen Cao put it, to Confucian values thousands of years old—embattled with a New Materialism that, by its nature, values the accumulation of wealth and property over the health of the natural world. I saw a coal industry whose inertia will take decades to slow or turn, confronted with the rapid rise of a clean energy sector that already dwarfs any in the West. There's plenty to get pessimistic about in China: a ravaged environment, a government determined to preserve its power at virtually any cost, an economy hurtling toward what very well could be an implosion to make the 2008–09 financial crisis look like a minor slowdown. And yet I came back strangely, cautiously optimistic. In part, this was surely wishful thinking; but there are at least three reasons to believe that China, more than the liberal democracies of the West, has the will and the capacity to meet the overwhelming challenge of climate change.

The first is China's dynamism. The sheer, overwhelming vitality of the place, the impressive feats of public infrastructure and education, the sleepless, striving energy of the people, the strong stirring of patriotism and pride, the awakening sense of a shared commitment to a clean environment—all of these strike jaded visitors from the late stages of American empire as things we have lost and don't know how to regain. It's hard to travel in China without concluding that this is a people capable of accomplishing whatever needs to be accomplished.

That native energy is complemented by an uncomfortable truth: it's easier, in China, to accomplish big things, particularly massive infrastructure projects, than it is in the democratic, disputatious West. The strong men in Beijing may have trouble reining in the coal industry, but they displaced millions of people to build the huge Three Gorges Dam on the Yangtze. It also helps to have something on the order of $1 trillion in capital reserves. If necessary, the Chinese government can harness deep wells of capital and compliance—or at least acquiescence—from the citizenry that Washington can only dream of.

The second factor is, of all things, bilateral cooperation, particularly with the United States. The United States and China clash over a long list of issues that includes intellectual property rights, the contested islands in the South China Sea, trade conflicts, and, of course, human rights. Until the historic Obama-Xi climate change accord, in November 2014, that list also included internationally enforceable limits on carbon emissions. But climate change is one area where the interests of the two countries overlap and, in many ways, converge, and there are growing signals that mutual cooperation, technological and scientific exchanges, and high-level policy talks will move the two countries forward in ways that overarching global agreements could not.

One example is the development of liquid-fuel thorium reactors, which I wrote about in my previous book, *SuperFuel*. With expert collaboration from the U.S. Department of Energy, China's National Academy of Sciences is pushing forward with an ambitious and well-funded program to develop these reactors, which promise an abundant source of clean, safe, essentially inexhaustible energy. It's entirely likely that China will build a prototype liquid-fuel thorium reactor by 2016. Offering a quickly scalable alternative to coal-fired power generation, that would be a landmark in the struggle against climate change.

The final reason for optimism is symbolized by the piece of Chinese calligraphy I keep on my wall in my office in Boulder: the characters *wei* and *ji*, popularly translated to mean "danger" and "opportunity."[24] "Crisis equals risk plus opportunity," it's said. People like Liang Sheng Cai, the elderly ex-miner at Qing Ci Yao, have ridden out upheavals and revolutions on a scale that, for most Americans, are experienced only as tales told by their grandparents.

Within the lifetimes of those now living, the Chinese have remade their society multiple times. Living on a hotter, drier planet will require similar adaptations, and China will be affected more than almost any other country. In the spreading chaos of global warming and the splintering of the *Pax Americana,* hundreds of millions of Chinese people will spy opportunity. Conquering coal, and averting climate catastrophe, will require economic, social, and political transformations of the highest order. Those are Chinese specialties.

PART IV
GROUND ZERO

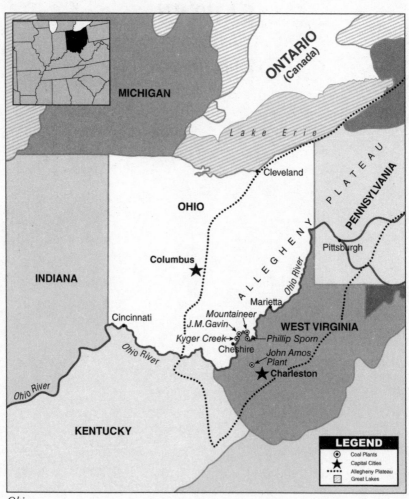

Ohio

CHAPTER 9

OHIO

Scotty Lucas was the former mayor of a town that no longer exists. This double obsolescence seemed to faze him little, which was not all that surprising considering that he had outlived his wife, one of his children, and the town he'd spent most of his 81 years in.

Lucas's one-story brick home, with a bass boat in the driveway and wrought iron patio furniture in the carport, is one of the few still standing in Cheshire, Ohio. This riverside village became briefly famous a dozen years ago, when American Electric Power (AEP), the utility that operates two large coal-fired power plants here, bought Cheshire, all of it, for $20 million rather than continue to deal with the complaints of the local citizens over the air pollution to which their 140-year-old community was exposed.

I visited with Lucas on a mild September afternoon as puffy white clouds, overhanging the Ohio River, melded with the smoke and steam billowing from the smokestacks and cooling towers of the Gavin Plant, built in the early 1970s, the largest coal plant in Ohio and one of the largest in the United States. Just downriver was the smaller, older Kyger Creek plant, which has been burning coal to make electricity since 1954. Lucas, a hospital administrator, served as Cheshire's part-time mayor from 1970 to 1998. He kept getting reelected, he told me, because "nobody else wanted to catch the flack." His successor, Tom Reese, helped negotiate the $20 million buyout that absolved the utility from any future liability for damage to the locals' health or their property. All but a handful of the town's 450 or so residents accepted the buyout; those above a certain age, like Lucas and his wife, were allowed to remain in their homes for the rest of their lives.

Was the buyout a good thing for the town and its residents? I asked. Lucas paused for nearly a minute. "We were given a lifetime estate. My wife was ill, she didn't want to relocate. It was okay for our particular needs."

Lucas's wife died "inch by inch," as he put it, of pulmonary fibrosis, finally succumbing in 2012. I asked him if he thought her illness was caused by their proximity to the coal plants. "That question arose," he said. "The doctors, they wouldn't comment." And what does he think?

"Naturally, it lights up. It certainly could be."

One of his three sons was a foreman at the Gavin Plant. Later on in our conversation he returned to the question of whether the buyout was a good thing or not. "I hate it that they uprooted that many people, especially in a place with a history like this village. It wasn't a good outcome. It's not very rewarding to look back on all the work we did. I can remember when we didn't have paved streets, didn't have gas service, didn't have running water. We did all that. We had three playgrounds, and good schools. It was a great place to raise a family. That's all gone."

The plants, though, were still there, consuming around 35,000 tons of coal every day. A gray mountain of coal ash loomed above what's left of the town. AEP, though, has not entirely escaped liability for the environmental effects of the plants: in early September 2014 a new lawsuit was filed on behalf of 77 contractors and families of contractors who worked at the coal ash land-fill. Seeking unspecified damages, the complaint claimed that workers were "exposed, unprotected, to coal-combustion-byproduct waste, a radioactive amalgam of hazardous constituents that pose known risks for human health."[1]

With no one left to buy out, AEP said it plans to fight the lawsuit.

Cheshire is about 90 miles southeast of Columbus, on the downstream leg of a huge S-bend in the Ohio River, on the back side of the Appalachians where Ohio, Kentucky, and West Virginia collide. This is the heartland of the coal industry: besides Gavin and Kyger Creek, there are three other coal-fired power stations within a 50-mile radius: Mountaineer, Philip Sporn, and John Amos, all in West Virginia. Barges have carried coal down the Ohio for more than a century, and coal mines and power plants have been the major source of employment in the region since World War II. To say the least, this is not a hotbed of anti-coal activism.

But in 2000, when a bluish, sulfurous fog started blanketing the town on some days, bringing headaches, scratchy eyes, and sore throats, the townspeople started to complain. Monthly meetings between Mayor Reese, the town council, and AEP officials became acrimonious. The town hired a Washington law firm to represent it, and talk of a class action lawsuit over the "blue plume," as it was known, became common.

At the same time, AEP was coming under increasing scrutiny from federal officials. The Environmental Protection Agency declared Gavin in violation of the Clean Air Act in 2000. A report from the Centers for Disease Control and Prevention found sulfur compounds in the air in Cheshire that were five times over the level needed to trigger an asthma attack. Its options narrowing, the utility decided to write a check. Actually, a bunch of checks. Each homeowner inside the town line would get three and a half times the appraised value of his or her home. In return, they would vacate and sign an agreement giving up not only their rights to sue AEP, but also their heirs' rights. The elderly and the infirm could sell out and stay in their homes until they died; everyone else who sold had to move. The Columbus-based corporation was now the owner of a town, which it bulldozed, plowed over, and replanted. Scotty Lucas's house sat amid a parklike swath of green grass dotted with trees. It was hard to tell there were once houses on each lot. In fact it's hard to see that there were lots. The old Cheshire school, a brick rectangle with tall arched windows, sat mostly empty. Cheshire Baptist Church declined to sell, and still drew worshippers from across Gallia County to Sunday services.

The origins of this extraordinary deal—the question of who actually came up with the buyout proposal—have grown over with myth and countermyth. Some say the utility came up with the scheme to silence its local critics. Some say it was the mayor and the city council who sold out the town for what they thought would be a personal windfall. Officially, AEP made the $20 million deal not to avoid pollution lawsuits, but for "future expansion"—which has never occurred. As for Mayor Reese and the town fathers, they had two bad options: lawyer up, and spend the next ten years or so trying to fight the coal company, or sell and abandon the town. A lot of the townspeople probably wouldn't live another ten years. Reese said several times that the end of Cheshire was a sad but unavoidable development—and that no personal gain was involved. "We want to make sure that what we do is legal and proper," he told the Associated Press.[2]

All but a handful of them took the money. But the buyout divided neighbor from neighbor and, in at least one case, brother from brother.

"My brother sold out," Ron Cornelius told me. "It didn't help our relationship. We've mended it, though. We don't talk about it."

One of the few owners who refused to sell, Cornelius, lanky and voluble at 74, still lived in a gracious two-story brick colonial with a small swimming pool out back. Across the street was another boarded-up school. Immediately adjoining his property was a fenced-off area of dense shrubbery that hid a gravel pit that feeds the coal plants. Cornelius and his younger brother grew up playing in the river and the surrounding hills. Twelve years after the buyout, he didn't bother to conceal his bitterness.

"It really hurt me," he said. "I was a trustee of the church. I spent 15 years gone from here, and I didn't come back home to get pushed around by the company. I grew up looking at these hills. To wake up and not see them . . . you might as well put me in the hospital.

"Now the town's gone but I'm still here. I've got a place out at the cemetery, and I'll be there one day."

The story of Cheshire will receive renewed attention in 2015, with the new lawsuit and the release of *Cheshire, Ohio,* a documentary by Brooklyn-based filmmaker Eve Morgenstern, in postproduction when I visited southern Ohio. Morgenstern, who has been collecting footage and interviews in Cheshire for nearly a decade, had seen her own feelings about the buyout grow more complicated over time.

"For those who believed in the buyout as a solution it was a good thing," she told me in an email.

> It allowed people to sell their homes for a decent price that they might not have been able to get otherwise. It allowed them to get away from the pollution problems and alleviate the daily stress of worrying about the next problem or the health effects. For those who were against the buyout it was a bad thing. It didn't force the plant to clean up the pollution and it eliminated a

town and a community that was very beloved, where people had long histories and very strong emotional ties.

When I started filming, I felt the buyout was a bad thing. I could not believe that an enormous and profitable energy company like American Electric Power, located precariously close to residents, could not be held more accountable. But after hearing about the long, exhausting struggle of the residents and learning that the plant was clearly never going to clean up the pollution (thanks also to the support for energy companies in this country and to the problematic state and federal EPA regulatory process), I understood why residents took the buyout and saw the value in their ability to get out of Cheshire and away from the plant.

We've all seen how the energy industry has wreaked havoc on communities from fracking, chemical spills, and harmful emissions from coal-fired power plants. At the same time we are all complicit in our dependence and use of coal as an energy source and that's what has led to climate change, one of the biggest tragedies of our times.[3]

We are all complicit. This might sound like a black-and-white story: the evil coal company pushing out the simple, sturdy townsfolk. But it's not that simple. It's clear that, whoever actually first uttered the word *buyout,* the town fathers knew a chance to cash out when they saw it. Terry Beebee, another retiree, lived up Highway 7 from the plant, on the wrong side of the road: the town limit lay just beyond his front yard, and those outside the town proper (known as a corporation in Ohio) were never offered buyouts.

"The Cheshire people got the money, and those of us who didn't live in the corporation got nothing," he told me. He walked me to his white car and ran his finger across the hood. "See that?" I did the same, and my fingertip came away with a fluffy coating of light-gray coal dust.

It's also clear that not everyone resents the way things turned out.

"You can't beat it," said Jim Rife, who grew up in Cheshire and spent 20 years living in Manassas, Virginia, before returning just as the buyout was happening. Like Cornelius, Rife, now 74, refused to sell. His white frame house sat on the bank above the river. His mother Gladys, now in her nineties, lived in the house opposite his. A flight of rickety steps led down to a homemade dock, where Rife's fishing boat was tied up. Few days passed when he was not on the river. "It's better than it ever was. You're out here by yourself, you ain't got neighbors around you making noise. If I want to take my rifle out there to the river and shoot it, there's nobody yelling at me. I've been real pleased with it."

For its part, AEP made an entirely defensible risk-management calculation: pay $20 million upfront to eliminate the possibility of years of litigation, and hundreds of millions in damages, down the road. Not many coal plants are built right next to towns anymore, and not many people have any idea where their electricity comes from, much less wake up every morning breathing the smoke from the stacks. AEP just happened to have a few hundred people, many of whose families had lived in the area for generations, living just outside the plant gates. No CEO would be fired for making that deal. What happened in Cheshire was capitalism in its purest form.

By late 2014, though, it appeared that a few of the old-timers might actually outlive the plants. Driven by dwindling coal supplies, abundant supplies of cheap natural gas, and proposed EPA regulations on power plant emissions of carbon dioxide, coal plant retirements were accelerating around the country. AEP has spent millions in recent years on smokestack scrubbers and other updated pollution controls at Gavin and Kyger Creek—an investment that could prove to be good money after bad. A 2009 study by the Institute of Southern Studies ranked Kyger Creek the twenty-ninth worst polluting coal plant in America.[4] The dams that contain its coal ash ponds are rated as "high hazards" by the EPA.

And even AEP, the largest owner of coal-fired power stations in the country, has taken dramatic steps to shift its generation fleet off of coal. Before the new EPA carbon regulations were announced, the company already had plans to retire 6,600 megawatts of coal-fired capacity over the next ten years. It's hard to imagine that Kyger Creek, and maybe Gavin, won't be among those. As electricity from renewable energy and natural gas spread, and the economics of coal deteriorated, utility executives like AEP CEO Nick Akins were facing the transformation of a business that has operated in more or less the same fashion for a century. Big coal-plant owners like AEP have to adapt or die.

Meanwhile, Cheshire's long fade into history continued. When Scotty Lucas dies, his place will revert to the company. When Ron Cornelius and Terry Beebee and Jim Rife pass on, their kids might move in, or, more likely, they might not. The pastor of the Cheshire Baptist Church was leaving for another church in a more populous town, and it was not clear how they would attract a new minister to a ghost town in a forgotten corner of southern Ohio. There wasn't much to draw anyone to Cheshire now that there was no school and no grocery and just one gas station/convenience store that served the plant workers heading home after their shifts. There's a good chance that within 20 years the few remaining houses will be empty and the plants shuttered.

Ironically, once the plants close, Cheshire will enjoy a strange sort of afterlife: decommissioning and cleaning up after coal plants requires men and machinery. Just ask American Municipal Power of Ohio, which announced in May 2010 that it would retire the Richard Gorsuch Generating Station, near Marietta, by the end of 2012. The Gorsuch plant opened in 1951. Charged with violating the Clean Air Act, American Municipal Power reached a settlement with the EPA and the Justice Department to shutter the plant and to spend $15 million on environmental mitigation at the site. The plant actually shut down in December 2010; demolition began in 2013 and, along with the cleanup, is expected to take much of the decade. American Electric Power's compliance plan—which includes not only shuttering more than 6,000 megawatts of coal capacity but also refiring another 1,070 megawatts with natural gas and building 1,220 megawatts of natural gas–fueled generation—is expected to cost $6 to $8 billion and last well into the next decade. Shutting down coal plants is a messy and time-consuming business, and much of those costs will eventually be passed along to ratepayers. In the interim, the shutdowns of the two Cheshire plants, when they occur, will bring another coal boom to the area, this time in reverse, as wrecking balls and bulldozers deconstruct the boilers and the buildings. That would bring a few years of jobs and activity to Cheshire. But the deconstruction crews are not going to move in for good. And they're not going to halt Cheshire's slow vanishing.

FORESEEING THE COSTS of decommissioning multiple coal plants, AEP, like many utilities, was playing a delaying game. In December 2013, the company proposed in a filing with the state utilities commission that it be allowed to charge consumers for the cost of operating the Kyger Creek plant even if the operating costs exceed the market value of the electricity it sells. In October 2014 the company asked for similar arrangements for four additional Ohio coal plants. Major coal operators FirstEnergy (which is based in Akron) and Duke Energy (which operates 15 coal plants across five states) have asked for similar deals. Essentially they would leave consumers on the hook for the difference in price between those megawatt-hours and the price of power on the competitive market.

Ohio first began to deregulate its utilities in 1999. In the wake of deregulation, both FirstEnergy and AEP spun off their generation and transmission entities into unregulated subsidiaries. The distribution utilities, which actually deliver power to customers, remained under the oversight of the public utilities commission. Essentially the utilities are seeking to have their regulated arms subsidize the unregulated ones.

"The utility companies are coming forward and asking for bailouts for these coal plants," said Sierra Club Ohio energy director Dan Sawmiller. "The message is, we're going to keep these coal plants open whatever it takes, so they're putting this fear out there about price volatility. They're the same ones that just a few years ago championed the open market, and now they're finding out that these old coal plants can't compete in this sort of market."

"The confluence of deregulation, flawed capacity markets and increasingly onerous environmental regulations is significantly changing the generation landscape in Ohio," said Pablo Vegas, president and chief operating officer of AEP Ohio, in written testimony before the Public Utilities Commission of Ohio. The payback scheme, Vegas declared, would "increase price stability to its customers and protect the local Ohio economies which it serves."

Translated into plain English, this means that Vegas is agreeing with Sawmiller: the old coal plants can't compete under the new regulations. But shutting them down, the utilities claim, would cause power shortages, eliminate hundreds of jobs, and decimate local economies. So they're asking customers to subsidize them, rather than making investments in the power plants of the future.

"They are trying to force these monopoly utility customers to pay for things that should have been left to market forces," Sawmiller told me. "It seems to me an attempt to preserve these coal assets. I don't know why you'd do that."

Actually, the reason is clear: AEP's earnings increased from $3.6 billion in 2013 to $4 billion in 2014, and its earnings per share went up 16 percent.

In the four years from October 2010 to October 2014, the company's share price climbed by more than 50 percent.[5] The status quo is working for Ohio utilities. Investing in the future of energy is risky, expensive, and complicated. Wall Street may have grown disenchanted with coal miners, but in 2014 it was still rewarding coal burners.

"In a deregulated market, you have to have price signals from the market that say 'We need to build'—there has to be a demand that's created and you have to support that demand," Nick Akins told *Columbus CEO* magazine in late 2013. "If you build a generation facility, it's in the billions of dollars. So you're making a pretty heavy investment and you need that price certainty around that. In the other states we have that. In the deregulated states we don't, at this point."[6]

On its face that's a reasonable argument, and predicting demand for electricity, since the 2008 financial crash, has proved to be a dicey proposition. What Akins is really saying, though, is the same thing his AEP Ohio counterpart, Vegas, said in his testimony: *We can't compete on an open market. We need guaranteed demand and guaranteed rates in order to survive.*

When the U.S. government bailed out the auto industry, it imposed certain conditions, including the resignations of top executives such as General Motors CEO Rick Wagoner, and new investments to make the automakers more competitive on the open market. Now, states like Ohio are being asked to bail out the big utilities by guaranteeing profits from outdated coal plants. It's no longer just the Greenpeaces of the world who are asking, why shouldn't the states impose similar demands?

SAM RANDAZZO LOOKED LIKE A MAN who would be at home in a back parlor in Las Vegas, with a cigar in his mouth, squinting at an unpromising hand of cards. Stubby, smoothly coiffed, and mustachioed, he had a gravel voice and a switchblade mind. He grew up in northern Ohio and worked his way through college at the University of Akron installing underground utility cables and pipes. He got his law degree at Capital University, a Lutheran school in suburban Columbus. By 2014 he was a lawyer at McNees Wallace & Nurick, a Columbus-based firm with big corporate clients in telecommunications, manufacturing, and energy. The former associate attorney general for the state of

Ohio, and former counsel to the state's public utilities commission, he has for two decades been the general counsel to the Industrial Energy Users of Ohio, a trade group devoted to keeping energy prices low and supplies abundant for big industrial users.

Early on, Randazzo knew that his future lay not as a politician but as a consigliere, a trusted adviser to politicians. He is one of those guys who crowd the background at bill signings and plant openings, while, up front, the governor or the senator beams and shakes hands. According to his official bio, he "has spent the last five decades solving issues affecting the price and availability of communications, natural gas, electricity and other regulated services."[7] Among his most memorable solutions was the key role he played in rolling back the renewable portfolio standards—the state-mandated requirements that utilities produce, and businesses obtain, a certain percentage of their energy from renewable sources—in Ohio.

"From an end-user perspective, the larger commercial and industrial users, most of those customers are doing things themselves to figure out how to get the most bang out of their energy buck," Randazzo told me the first time I talked to him, by phone from his office in Columbus. "The notion that there should be this one-size-fits-all imperative from the government—this much renewable, this much energy efficiency and demand response—that whole notion just gets in the way of how business thinks and acts. It locks out the opportunity for innovation and customization. There was a broad consensus among the business customers, on the consuming side of the meter, against these mandates."

That is not quite true. In fact, the Ohio Manufacturers Association, which includes major corporations with big factories in the state, including Honda and Anheuser-Busch and Proctor & Gamble, publicly opposed Senate Bill 310, which froze Ohio's renewable energy mandates, at least temporarily. But Randazzo came off as a man unhindered by doubt or nuance, and his position—that the only way to avoid economic disruption, if not outright collapse, is to preserve the status quo in the energy sector—is the default argument of everyone who supports propping up monopolistic utilities and keeping aging coal plants running.

In 2008 Ohio's legislature passed almost unanimously one of the most ambitious renewable energy laws in the United States. The original legislation required utilities to get 25 percent of their power from renewable and other clean energy sources by 2025, while cutting their customers' power usage by 22 percent. In 2013 Ohio got just 1 percent of its total energy from renewables, according to the EPA.[8] In June 2014, Ohio, which gets 70 percent of its electricity from coal-fired power plants, became the first state in the United States to roll back those clean energy mandates with the passage of SB 310.[9] Coming just a week after the EPA announced its tough new rules on carbon emissions from power plants, the move threw into question the future of government-supported renewable energy markets and brought a respite to the two major utilities headquartered in the state, AEP and FirstEnergy of Akron.

Environmentalists and renewable power providers fear that Ohio could be the spearhead for pushing back renewable mandates in other states. The complicated politics, the presence of two of the nation's largest utilities, the legacy coal industry (Ohio produced 26,000 tons of coal in 2012), and a healthy record of clean-energy investment over the last five years—now endangered—made Ohio the primary battleground for the struggle between clean energy and cheap, dirty coal. "This is ground zero," Dan Sawmiller told me.

The fight over SB 310, carried out over the first six months of 2014, was legislative sausage-making at its least edifying. The bill was supported by the big utilities, Peabody Energy, and other coal companies, along with outside interests, particularly the American Legislative Exchange Council, which was funded by the Koch brothers and which poured millions into alarmist TV ads predicting calamity if the mandates stayed in place.

"Between 2009 and 2014 there was an accumulation of frustration, even hostility," Randazzo told me, "for how the mandates were working against the larger objective: to maintain employment, to grow employment, to do things that allow the larger customers to compete in the global economy."

This contention—that the coal industry and big utilities are a sort of full-employment tool for big industry, which can't compete otherwise—is unsupported by data. Under the current model, manufacturing employment has declined in the last two decades in Ohio just as it has nationwide (the last few years have seen a modest rebound, a development I'll return to later). Utilities, meanwhile, have struggled to compete in a semi-deregulated environment. Forced to compete for the first time in a relatively open market, the two powerful utilities "have taken to fighting each other like children," reported *Crain's*

Cleveland Business magazine in 2012, "engaged in a very adult fight over millions—perhaps billions—of dollars."[10]

Against this contentious backdrop, SB 310 moved toward passage. At
its root, this was a single-state election over a universal question: How much
should the government intervene in energy markets to drive society toward
cleaner resources and away from climatic disaster? That made it easy for opponents of the renewable mandates to cast it as American free market enterprise
versus socialized energy.

"I believe in an all-in energy market basket," Keith Faber, the president
of the majority Republican state senate, told me. "But anything you put in
that basket needs to have an independent, long-term viable strategy that's not
dependent on taxpayer or ratepayer subsidies. If the government is required to
subsidize any kind of energy in perpetuity, that's probably not a viable business
model."

This, of course, ignores the decades-old subsidies for coal, which have
slowed but not halted in the last decade. Indeed, the supposedly anti-coal
Obama administration had bumped up coal subsidies under the stimulus bill
that followed the financial crash of 2008, handing out federally subsidized
taxable bond funding to power companies looking to build new coal plants.
Among the utilities that took advantage was Ohio-based American Municipal
Power, which issued half a billion dollars' worth of these "Build American
Bonds" to help finance the Prairie State Energy Campus in neighboring Illinois. Completed in 2013, Prairie State includes a 1,600-megawatt coal-fired
power station—one of the largest coal plants to come online in decades, and
undoubtedly one of the last.[11]

For anyone who might be unconvinced by the socialized-energy line, Faber and his colleagues painted a bleak picture if SB 310 failed and the mandates held. The price of electricity would climb by 10, 15, even 20 percent a
year. Big companies would flee the state; thousands of jobs would vaporize;
the Rust Belt would fall further into decrepitude. The reality is that, even after
six years of the supposedly punishing renewable mandates, Ohio's electricity
rates (13.44 cents per kilowatt-hour for residential customers in July 2014, according to the U.S. Energy Information Administration, and 9.96 for commercial users[12]) remained among the lowest in the United States. According
to Susanne Buckley, a former AEP executive now a partner in Scioto Energy,

which consults with large electricity users, higher prices for power were coming for Ohio commercial users, but it wasn't because of renewable mandates. It was because the wholesale price of electricity across the region, which is set at auction, was about to soar—partly driven by retirements of aging coal plants. The regional capacity markets that conduct the wholesale auctions have their critics, but no one claims that they serve the interests of socialized energy.

And the mandates brought economic benefits of their own. According to the utilities' own reports, filed with the utilities commission, customers of FirstEnergy had saved more than $316 million on their electricity bills since the mandates were passed, resulting from $159 million in energy efficiency investments by the company. AEP spent $158 million on efficiency upgrades, saving its customers $366.95 million.[13] Trish Demeter, managing director of energy and clean air programs for the Ohio Environmental Council, testified before the legislature that overall, "Ohio's four investor-owned electric utilities have collectively spent $456 million on complying with Ohio's efficiency standard, saving their customers $1.03 billion to date on their electric bills."[14]

Ultimately, though, the renewable mandates were doomed in the state legislature, and Governor John Kasich, a moderate Republican who had supported the mandates when they were originally passed but who was facing a fierce reelection campaign, was backed into signing SB 310. Sam Randazzo couldn't help gloating.

"The opponents of the legislation viewed this as, 'We have to win Ohio, if we lose Ohio other states might get the idea they can do something too. We need to do any thing to keep from happening.' They had billboards all over downtown Columbus, all sorts of radio. They overplayed their hand frankly."

In fact it may have been the opponents of alternative energy, clinging desperately to the foundering coal industry in an increasingly stormy sea, who overplayed their hand in Ohio.

"I believe that Senate Bill 310 is the worst, the most self-destructive bill that I've voted on in my eight years in the General Assembly," State Representative Mike Foley, of Cleveland, told the Cleveland *Plain Dealer*.[15]

"At the end of the day the senate leaders had to break legs to get 310 passed," Ted Ford, the CEO of Advanced Energy Economy Ohio, told me. AEE Ohio is the state-level chapter of Advanced Energy Economy, the nonprofit business association created by billionaire climate activist Tom Steyer.[16]

"There's a standard of truth around some of this stuff that's going to come back and bite the other side. We had 21 large companies that came out in opposition to the bill, mainly because of the energy efficiency components [of the mandates]. They included some household names, like Honeywell, BASF, Honda, and Johnson Controls. That's a significant development as well."

If the Ohio legislation was a victory for the forces of reaction, it will probably one day be seen as a rearguard action, a sort of Battle of the Bulge in the larger coal wars. There's ample evidence that, faced with a choice between sticking with coal or paying slightly more for cleaner energy, both consumers and businesses will opt for clean energy. The coal industry and big utilities and their political agents in Ohio have only delayed the inevitable, and have probably harmed the state's long-term economic prospects while claiming to defend them.

"We had $12 million in signed contracts that went up in smoke" after 310 passed, Eric Zimmer, the founder and CEO of Tipping Point Renewable Energy, a solar project developer, told me. New solar power projects approved in Ohio in the first quarter of 2014, as it became apparent that the rollback of renewable mandates was likely to pass, dropped to their lowest level since early 2010.

Wind power suffered, too: EverPower Wind Holdings, a developer of wind farms, applied for a delay in launching construction of the Buckeye Wind Project, which would bring millions of dollars in investment to Champaign County in central Ohio. Up to ten pending wind power projects in the state, worth a total of $2.5 billion, could be stalled or killed by the passage of SB 310.[17]

Zimmer, whose solar company is based in Dublin, Ohio, is retreating for now, shifting Tipping Point's business model toward consulting and looking for opportunities in other states. What's happened in Ohio, though, is not the turning of the tide for coal; it's just a delaying of the inevitable. "Long term, the genie is out of the bottle," he told me. "They can slow it down in Ohio, by pausing the mandates. But the renewable energy industry is too big now, and the economics are too good, to kill it."

"What we saw here was a very strong concerted effort on the part of coal-fired generation supporters to protect the status quo," Ted Ford said. "But they're coming up against these strong national and global trends around innovation in energy that are unstoppable. Coal is an inefficient, dirty, costly way of generating electricity, and to protect that against all this new stuff . . .

"Look what happened in telecommunications. Any industry that depends on preserving a basic system that's not evolving as quickly as the rest of the world faces real challenges."

WESTERN OHIO IS HONDA COUNTRY. Since 1982, when the first auto production plant went in at Marysville (about 33 miles northwest of Columbus on Highway 33), the Japanese automaker has built a web of facilities that have transformed this bucolic area: the gigantic plant at Marysville (one of the largest manufacturing facilities under one roof in the United States), the Anna engine plant, a second full assembly plant at East Liberty, and the transmission plant at Russells Point. And in 2013, at Russells Point, Honda took another transformative step, installing a pair of wind turbines to feed electricity to the transmission plant.

I drove out from Columbus to see how Honda was adapting to the post-coal era. On a mild evening in September, the turbines, facing northwest, turned lazily in the breeze, bathed in the buttery sunset and casting long shadows across the adjoining cornfields. Owned by ConEdison Solutions and installed by Minnesota-based Juhl Energy, the 260-foot-high towers, each equipped with three blades more than half a football field long, supply about 10,000 megawatt-hours a year—about 10 percent of the plant's electricity. That's 10,000 megawatt-hours that won't have to come from the coal plant that supplies most of Russells Point's power.

These days there is nothing novel or startling about a pair of wind turbines standing in a Midwestern cornfield. What's more interesting than the fact of their existence is what they represent: auto manufacturing, one of the most energy-intensive businesses in the world, is gradually but decisively reducing its energy use and obtaining more of it from cleaner sources. Honda has trumpeted its "Green Factory" initiative, saying it plans to significantly reduce carbon dioxide emissions from its plants by 2020, compared to 2000 levels. By 2014, 10 of the company's 14 U.S. facilities produced zero waste that goes into landfills. Honda has made it clear that it wants to reduce its consumption of power from fossil fuels in order to be a better corporate citizen and to reap the accompanying good PR. But manufacturers of products from carburetors to catheters are finding that using energy more efficiently, and reducing their reliance on traditional, utility-centric, fossil-fuel-dominated energy resources, is good not only for the corporate image but also for the bottom line. Prodded by government regulations, shareholder unrest, and transformations in the energy sector, the manufacturing sector is transforming the way it sources and uses energy.

"Honda is always looking for opportunities to improve our efficiency and cost," Karen Heyob, the company's associate chief engineer in charge of energy strategy, told me. "I'm happy to say that we've kept energy usage stable this year despite significantly expanding our operations."

The energy revolution in manufacturing actually comprises twin revolutions. The best way to think of these changes is *extramural* and *intramural:* in other words, what happens outside the walls of the plant and what happens inside.

The external forces are well documented: falling prices for natural gas and renewable energy have made clean, distributed (i.e., located at or close to the plant) electricity generation a compelling business case. The effects of the shale gas bonanza on U.S. manufacturing are somewhat in dispute, but there is no question that low-cost natural gas is providing a competitive advantage to companies that make stuff in the United States—an advantage not outweighed by low labor costs in the developing world, particularly China. According to the U.S. manufacturing Purchasing Managers' Index, a measure developed by financial research firm Markit, manufacturing activity in the United States in September 2014 reached its highest point in more than four years. Factory employment, though still well below pre-2008 levels, is surging as well, reaching its highest level since March 2012.[18] A large part of this increase can be attributed to inexpensive natural gas.

The use of renewable energy in manufacturing, meanwhile, hovers around 8 to 9 percent, and it's not guaranteed to soar. A June 2014 study by the International Renewable Energy Agency found that that number could grow to more than one-quarter by 2030, or one-third if some form of carbon pricing takes effect—or, under current deployment plans and government policies, it could stagnate at 10 percent.[19] Those are global figures; the United States—with a robust clean energy technology industry, accelerating coal plant retirements, and tightening government regulations of carbon emissions from power generation—should see increases at the high end of that scale.

Moving from centralized coal plants to small, distributed power generation from renewable sources benefits manufacturers because the user controls the assets and long-term prices are fixed. What's more, shifting to clean energy now mitigates the risk of future carbon taxes or other forms of carbon pricing, Barry McClelland, Honda's longtime energy and operations chief who retired in June 2014 but still heads the energy committee of the Ohio Manufacturers Association, told me.

"Honda's view is you've got to go with the change," said McClelland. The company will "absolutely be adding more onsite renewables in the future."

The benefits of alternative energy are not evenly distributed across all manufacturing locations, though. Honda's Ohio complex, for example, is contained within the service area of Buckeye Power, a rural electricity co-op that has been untouched by utility deregulation and which still gets the bulk of its power from coal. Honda had to reach a power purchase agreement with Buckeye in order to erect the Russells Point turbines. But Honda displayed its commitment to new energy in the political fight over Ohio's renewable portfolio standards—the state-mandated renewable energy requirements that were frozen, at least temporarily, by the state legislature in May. Honda was among the large manufacturers in Ohio that opposed the rollback. As McClelland says, you've got to go with the change.

WHAT'S HAPPENING INSIDE THE PLANT is more complicated but, ultimately, equally transformative. Simply put, a confluence of technology advances in networks, wireless sensors, virtualization, and monitoring equipment is enabling advances in manufacturing efficiency, energy conservation, and quality control that were hardly conceivable just a few years ago.

This shift is coming as a shock to old-line factory managers unused to calculating energy as a key metric of efficiency and productivity. "No one ever got fired for purchasing a pump or a machine that's too big for the job," said Fred Discenzo, manager of R&D at Rockwell Automation, at a recent energy management conference in Akron. In manufacturing, "excess capacity has always been the free option."[20]

Rockwell is among an emerging segment of technology vendors that is trying to change that, through what it calls "the connected enterprise." What that means is connecting the factory floor to the C-suite with far greater visibility and immediacy than before. Another name for this change might be "extreme granularity": in the near future, energy use will be measured not at the factory or line or machine level, but at the individual process level, per unit of production: how much energy did it take to make this widget or valve or bag of ice, and where in the process can that energy use be optimized?

Rockwell itself produces more than 300,000 products, mostly controllers, visualization panels, and other automation devices as well as the software to run them. By 2014, at its Twinsburg, Ohio, plant, the company was seeing the culmination of changes that have been decades in gestation. John Nisi, Rockwell's vice president of market development, called it "using controllers to make controllers."

"In that facility, similar to our other plants, it's taken us over eight years to connect the enterprise," Nisi told me. "A lot of manufacturers have been hearing us talk about this at different stages for 30 years. Only recently has the technology gotten to the point where it's really attainable."

Consider an oven that heats printed circuit boards. In a traditional plant, monitors ensure that the machine is kept at a consistent temperature, whether there's product moving through it or not. In Rockwell's new systems, a set of engineered algorithms tracks the movement of boards down the line and, if the oven's waiting for product, sends it into standby mode, cools the temperature down a few degrees, and then reheats to the optimum temperature in time for the next run of boards. Of such minute adjustments are major energy revolutions made.

These twinned revolutions—cleaner, cheaper, more distributed energy coming into the plant, and sophisticated automation technology reducing energy intensity inside the plant—will result in changes that have far-reaching implications for the manufacturing sector, and for the economy. At 32 percent of total energy consumption, industry uses more energy than any other sector of the U.S. economy.[21]

Manufacturers who adapt to the new realities of energy by changing the ways in which they source and use electricity will be more competitive on the global stage—and could help usher in the new economic upswing that politicians and analysts have been dreaming of for years.

That's because not only is the United States enjoying an energy bonanza thanks to low-cost shale gas, but the changes that Nisi described will help erode the low-cost labor advantage that has shifted much manufacturing to China and other developing economies, away from the developed world.

"More and more manufacturing operations are becoming less and less dependent on skilled labor," Nisi explained. "If you want to sell Oreos in China, then you make 'em in China. If you're going to produce cars, or other sophisticated products, you've got to balance the cost of materials and feedstock in the supply chain to produce them. As the middle class grows, and labor rates start to equalize, the costs of goods sold becomes more a factor of inventory and cost of goods to transport. Labor becomes less and less important."

"The new era of manufacturing will be marked by highly agile, networked enterprises that use information and analytics as skillfully as they employ talent and machinery to deliver products and services to diverse global markets," concluded a 2012 McKinsey study entitled "Manufacturing the Future."[22]

Agile, networked enterprises may need fewer workers. Already, they're using less energy. Total energy intensity in the U.S. manufacturing sector fell by 17 percent from 2002 to 2010, according to the Energy Information Administration.[23] Energy innovation equals agility. There's a subtle but increasingly pervasive link between companies' and economies' ability, and willingness, to shift away from traditional energy resources and market structure—i.e., coal—and their ability to compete in a high-tech globalized market.

NEEDLESS TO SAY, this realization has not set in everywhere. On my last day in Ohio I attended the Northern Ohio Energy Management Conference, put on by the Manufacturers' Education Council. Given the turmoil in energy markets in Ohio, I arrived expecting a balanced and forward-thinking discussion of the transitions manufacturers and energy providers are facing. That's not what I heard.

The keynote speaker was Jonathan Lesser, the president of Continental Economics in Sandia Park, New Mexico. For a sample of Lesser's thinking

I took a look at his blog, which included a column called "New Data May Indicate a Pause in Global Warming": "Although individuals who believe in human-caused climate change say 'the science is settled,' science is never settled. . . . The inability of current climate models to explain the lack of warming over the past 15 years, plus predictions that the sun may be entering a prolonged period of low activity, suggest that much more analysis must be done before we impoverish ourselves by choking off economic growth. Don't put that winter coat in storage quite yet."[24]

The Manufacturers' Education Council, in other words, had invited a climate change skeptic to keynote its energy management conference. In 2014. It was like inviting the president of the Flat Earth Society to kick off a geoscience conference. Lesser's talk was entitled "Ohio's Energy Renaissance."

"With the passage of SB 310, Ohio is moving forward with a renewed emphasis on market competition," Lesser declared, "one that takes advantage of Ohio's shale gas reserves and promotes new investment and recognizes the value of the reliability of existing baseload generation."[25]

In other words, long live the status quo. "This is the biggest group of energy dinosaurs you can imagine," an attendee muttered to me as the crowd applauded Lesser.

After Lesser came a panel discussion that included Sam Randazzo and William Ridmann, the vice president of regulatory affairs for FirstEnergy. Ridmann, as you might expect, delivered a forceful celebration of SB 310 and a repudiation of the proposed EPA limits on carbon emissions from power plants. On the EPA rule, Ridmann offered the classic denialist position: more study is needed. "The chances of getting this done in the time frame [proposed by the EPA] is a very aggressive assumption," he said. "They're asking states to essentially reengineer their energy systems. It's a monumental task—in the past rule-making took five years. This could take longer than that."

Time, of course, is exactly what we do not have. In November 2014, as I was making final edits to this manuscript, comments were filed with the Ohio Public Utilities Commission in support of FirstEnergy's petition to direct taxpayer subsidies to its unregulated generation unit, in order to keep the Sammis coal plant, on the Ohio River near Stratton, operating. The commenter argued, naturally, that jobs were at stake—jobs that Ohioans couldn't afford to lose. FirstEnergy says the Sammis coal plant employs 400 people and keeps another 1,600 working for suppliers, service providers, and contractors.

"Many Ohio coal mining and related (eleven to one multiplier) jobs depend on keeping this power plant in existence."[26]

The submitter of the comments was Robert Murray, the head of Murray Energy—a "key coal supplier" to First Energy.[27]

A week earlier the Intergovernmental Panel on Climate Change released its latest report on the extent and likely course of global warming. "Warming by the end of the 21st century will lead to high to very high risk of severe, widespread, and irreversible impacts globally," the report stated. The likely impacts before 2100 will include the collapse of the West Antarctic Ice Sheet, sea-level rises of three to four meters, severe widespread drought and desertification, more disastrous and more frequent extreme weather events, and the extinction of a large number of the species now living on the planet.[28]

The change was coming. The meteor was hurtling toward Earth. In Ohio, though, dinosaurs still walked the earth, grazing contentedly.

EPILOGUE

THE RUHR

Founded in 1898 as *Rheinisch-Westfälisches Elektrizitätswerk,* RWE is one of Europe's largest utilities and Germany's largest burner of coal. The company is based in Essen, in the valley of the Ruhr River, for centuries the center of iron and steel production, weapons manufacturing, and, naturally, coal in Germany. Among the most heavily industrialized places on earth, the Ruhr is the cornerstone of German industrial might. Its factories powered the unification of Prussia in the late nineteenth century, the buildup of the Wehrmacht in the twentieth, and the German economic resurgence after the fall of the Berlin Wall. RWE was created to provide electricity to the steel mills, gun factories, and chemical plants of the Ruhr, and for nearly a century it was reliably profitable. Until the last two years.

Like many European utilities, RWE has been buffeted by a series of headwinds that were unforeseeable only a few years ago. Germany's *Energiewende*—the "energy transformation"—established subsidies that pushed renewable energy to more than one-quarter of Germany's energy supply. On some sunny and windy days, when solar and wind power are abundant, big power suppliers like RWE have been forced to pay customers to take electricity off the grid. Before the financial crash, the company placed large bets on building new generation capacity, including the 1,100-megawatt Niederaussem coal plant, which won't begin producing power in 2018. Niederaussem will cost the company more than $2 billion to build. From the time that plant was greenlit until late 2014, the price of electricity in Germany crashed, making such large plant investments untenable. The price RWE gets for wholesale electricity fell by half, to around $54 a megawatt-hour, from 2000 through 2013.[1] How can you operate a coal fleet under these circumstances?

"The answer is that it's basically impossible," Thomas Birr, a senior executive with RWE, told *Mother Jones* magazine.[2] In 2013 RWE lost money for the first time since World War II.

The weak economy and the soft electricity market drove European utility stocks down 12 percent from 2009 to 2014, even as the overall market gained 60 percent, according to Bloomberg.[3] And they fueled a comeback for coal that no one anticipated when the *Energiewende* got under way.

After Chancellor Angela Merkel announced the phase-out of Germany's nuclear fleet in 2011, following the Fukushima nuclear disaster in Japan, coal use in Germany rose 13 percent in three years.[4] As the price of carbon permits, under the European Union's Emissions Trading Scheme, slid below $10 per ton, it was cheaper for utilities to burn coal and pay for the allowances than shift to renewables or other relatively high-cost fuels, such as natural gas. What's more, Germany's boilers burn mostly lignite coal, the soft, brown fuel that produces less energy and much more carbon emissions per BTU of energy than anthracite or bituminous coal.

The coal renaissance in Germany sparked fierce protests across the coal regions and questions from officials who wondered why one of Europe's greenest countries, the birthplace of the worldwide green movement, couldn't kick its coal habit more quickly. It also prompted plenty of schadenfreude among coal industry supporters and energy "realists" in the United States, offering "a cautionary tale in what happens when you try to substitute green dreams for economic realities," as the *Wall Street Journal* put it.[5]

Coal was on the march in other countries in Europe, including Poland and the United Kingdom. At least two dozen new coal plants are scheduled to start up in Europe in the next decade, according to the World Resources Institute.[6] To the surprise of officials and greens, Europe had entered "some kind of golden age of coal," Anne-Sophie Corbeau, a senior analyst at the International Energy Agency, told the *Economist*.[7]

There were signals that the coal binge in Germany could be a temporary aberration—for one thing, utilities are burning coal at aging plants as fast as they can before some of those plants become obsolete under new air pollution limits that will come into force in 2016—but as of late 2014, the *Energiewende* had brought a cascade of unforeseen consequences, most of them bad.

Geopolitics had not helped. Most of Germany's natural gas comes from Russian gas giant Gazprom, under contracts signed years ago that keep prices high. Germany, naturally, was eager to reduce its dependence on Russian gas, particularly after the invasion of Ukraine in 2014. It was hastening to develop domestic supplies of shale gas while rushing to build import terminals for U.S. gas—which, even with transport costs, could be inexpensive enough to compete with coal. Those developments, though, are likely years away. In the meantime, in the industrial heartland of Europe, King Coal was not close to abdicating.

AS I WRITE THIS, coal is on the march in Germany. Central China is enduring the greatest fossil fuel project under construction today, perhaps ever. In Ohio, they have rolled back renewable mandates; the Republican sweep of the 2014 midterm elections in the United States ensured that no national climate change legislation will be passed anytime soon; in the Powder River Basin, the miles-long trains arrive empty and leave full, 24 hours a day. The infernal machine clanks on, heedless of impending disaster. It's hard, in late 2014, to be optimistic about the future of coal, because coal still has a future.

Yet I am optimistic. In part this is in the way that a blind man is confident, not that he might one day regain his sight, but that the visual world, if he could see it, is more colorful and dazzling than the one in his imagination. I have a 14-year-old son. To believe that we will burn black rocks until we choke on the fumes would be to give up believing that he and his children will inhabit the rich and verdant world that I have loved. To do that would mean, not merely to be defeated, but to acknowledge defeat. No amount of staring at coal mines and touring coal-fired power plants could make me do that.

But there are rational reasons for optimism, as well. In the fall of 2014, evidence gathered that the financial markets were moving on from coal—that big financial institutions and private investors had started to factor climate change into their decisions. That would be a death blow that no EPA regulation could equal.

In May 2014 Stanford University's board of trustees voted to stop investing the school's endowment in coal mining companies. It was the most visible victory to date for the fossil fuel divestment movement that has rippled across campuses and cities in the United States and Europe in the last few years. But it overshadows a less heralded but deeper trend that gained momentum in 2014: traditional investors, including Wall Street equity firms, sovereign wealth funds, pension funds, and other institutional investors that control large amounts of capital, are quietly moving away from coal—driven not by their consciences but by their bottom lines.

These investment decisions are based on conventional, hard-eyed measures of risk and return—they're not "values-based" but value-based. And they signal a far more dangerous development for the coal industry than the growing distaste of liberal academia for dirty fuels. Follow the money, and the path suggests that coal is losing the battle for the future.

"Utilities and other electricity producers are poised to invest heavily in retrofitting their old plants or in building new ones," observed a report produced by the Union of Concerned Scientists in 2011, before the fossil fuel divestment movement had gotten traction. "Each major retrofit or new plant represents an enormous long-term financial commitment to coal power. But . . . current economic, technological, and policy trends make such commitments exceedingly risky."[8]

More recently, a May 2014 analysis from the Institute for Energy Economics and Financial Analysis, bluntly titled "NYC and NYS Pension Funds Should Divest Coal Stocks," put the case against coal succinctly: "The current position of the U.S. coal industry, and increasingly that of coal producers worldwide, is weak. And the worst is yet to come. U.S. coal company leadership has no effective investment rationale for improving stock performance . . . Selling the stock would actually put the money to more profitable use and better protect the beneficiaries of the funds."[9]

And coal carries its own set of risks that are different from, and in some ways greater than, those of oil and gas.

For one thing, coal mining, although it's a less capital-intensive business than drilling for oil, is a brute-force enterprise: there are no technological innovations on the horizon that will make coal extraction cheaper, or open up large new reserves that were previously inaccessible or uneconomic. You can't frack coal.

Also, coal, unlike oil, is replaceable. While there are few substitutes for petroleum in many transportation applications, such as aviation, substituting natural gas, or renewable energy, for coal in producing electricity is feasible and increasingly economical. And it's happening right now. What has shifted for the coal industry in the last several years resembles the public disaffection with tobacco a couple of decades earlier: once a ubiquitous part of modern life, it has become something to be avoided, even shunned. Once that happens, the contraction of the industry follows very quickly. Coal was your grandfather's fuel, and if you continue to put your money into it in the twenty-first century, your grandchildren are going to wonder why you did.

THAT BEING SO, I will end on a scene not of destruction but of rebirth. In July 2013, British prime minister David Cameron cut the ribbon on the latest redevelopment plan for the Battersea Power Station in London. The Art Deco monument, with its four tall chimneystacks set against the south London sky, had essentially been abandoned for three decades.

"This is a great day," said a beaming Cameron. "It has been a long time to get the Battersea Power Station development going." London mayor Boris Johnson said the Battersea renovation "is now sparking the wider rejuvenation of a once neglected part of London into a vibrant new quarter."[10]

Battersea became operational in 1933, with a second unit added in 1953. Battersea A, the original boiler, stopped producing electricity in 1975, and Battersea B was shut down in 1983, just before the coal miners' strike that led to the destruction of Britain's mining unions and the long, slow decline of the British coal industry, which had launched the coal age in the mid-nineteenth century. The plant had achieved an odd afterlife in pop culture: it was a primary backdrop for the Beatles' 1965 film *Help!,* and it appeared on the cover of Pink Floyd's 1977 album *Animals.*

Besides its evident popularity with rock bands, Battersea was considered a prime candidate for redevelopment. An initial plan to create a large amusement park collapsed. An Irish real estate developer bought the site for $607 million in 2006, planning to turn it into a luxury residential and shopping complex. That scheme was undone by mounting debts. At one point the owners of Chelsea Football Club proposed to build a stadium at the power station. Battersea's status as a historical building means that any new project had to preserve the ornate brick building, making redevelopment an expensive prospect. For years it seemed that Battersea would remain a beloved but abandoned hulk on the bank of the Thames.

In partnership with Asian conglomerate Sime Darby, a Malaysian development firm called SP Setia bought the property in 2012 and announced a plan to build a massive residential and shopping complex with nearly 900 luxury apartments. The developers were reportedly in talks with Emaar Properties of Dubai (builder of the Burj Khalifa, the world's tallest building) to open a five-star hotel at the site. Suddenly it appeared that Battersea could become a symbol of the transformation and rise, literally from the ashes, of old coal plants. That the furnaces could become gardens.

But Battersea's long abandonment also illustrates the challenges of cleaning up and repurposing retired coal plants—of which there are going to be hundreds in the coming years. Market forces are going to kill off coal, but market forces are corrosive, not explosive. And we need to blow the plants up. Blowing things up requires tough choices and meticulous preparation. If the coal industry dies in agony, like a crippled, flailing leviathan, it will take our cities with it. If we murder it swiftly, its passing could be less destructive. In the two years I worked on this book people often asked me, "What's your solution?" I don't have a solution, but I did arrive at three principles from a which a solution, or a set of solutions, could emerge.

The first is, **All of the above won't work.** Not wanting to alienate the fossil fuels industry, politicians and industry executives are fond of saying "We need an 'all of the above' energy strategy." Coal, renewables, nuclear, natural gas: throw it all in a basket and we will develop all of them simultaneously, building our way out of the trap we've fallen into.

That's like an obese, three-pack-a-day alcoholic saying "I'll take an all-of-the-above approach to healthy living—I'll keep smoking, drinking, and eating fatty foods, but I'll also eat plenty of fruit and vegetables." A sustainable energy strategy requires making choices, and the first and most crucial choice is to shut down coal.

Second, **We can't abandon the workers.** There has been talk, in the last few years, of a "just transition"—a plan for retraining, reeducating, and repurposing coal miners for more modern tasks, like installing solar panels.

Unfortunately, most of it is still talk. The people of Holmes Mill and Craig have been abandoned. There is no economic rebound on the horizon in Appalachia and the other coal mining regions of the United States, or Europe, or China—as Terry Sammons told me, "Google is not moving to eastern Kentucky." There is plenty of evidence that moving off coal, and toward the new energy economy, will be beneficial to the economy as a whole; but that won't help the miners who have been making $60,000 or $70,000 a year with only high school educations and without other marketable skills. Leaving those people to their own devices and the McJobs economy is not just morally indefensible; it's bad public policy.

There are few precedents for supporting and re-equipping whole industrial populations for new jobs. Britain's Industrial Training Act of 1964 is today considered a failure. Created in 2010, the Appalachian Regional Development Initiative is another in a long line of noble but ineffectual bodies formed to bring new ways of making a living to the people of the mountains. Another high-minded development initiative is not what former coal miners need; they need access to education and training and a way to support themselves while they get it. As it happens, there *is* a precedent for a massive federal aid program to feed and support and educate a large number of workers made obsolete by history: the G.I. Bill.

I am only half joking. World War II veterans had risked their lives to safeguard the free world and end fascism. Coal workers have for decades risked their lives to bring light and heat and power to our houses and factories and office buildings. The country had a moral obligation to its returning soldiers, but it had a political and economic incentive to get them to work, too. Just as no rational government would want a large population of well-trained, alienated, and idle soldiers on its hands, who wants a large number of well-armed, idle coal miners sitting around growing angrier and more disaffected? That's a recipe for economic stagnation, antigovernment radicalism, and civil unrest.

How much would it cost to pension off the entire coal work force, tomorrow, and provide a serious, free, comprehensive education and retraining program for all who wanted it? The Energy Information Administration estimates that just under 1 billion tons of coal were produced in the United States in 2013.[11] At a surcharge of one dollar per ton, that's $1 billion a year. According to the Bureau of Labor Statistics, there were slightly more than 80,000 people employed in the coal industry as of 2013.[12] The National Mining Association, a trade group that includes "support services" in its tally, counts more than 195,000. Weighting the federal estimate more heavily, let's round up to 100,000. That's $100,000 a year for every coal worker in America. That's enough to make the payments on your pickup truck and pay for tuition at the local community college.[13]

Finally, **The solution must be global.** Climate change is the ultimate transnational effect, a working out on a planetary scale of the tragedy of the commons. The coal industry is fully globalized. It will do no good to shut down the mines of the Powder River Basin, or the coal units of American Electric Power, while doing nothing about Shenhua or Tong Mei. International talks on climate change have gone mostly nowhere, but that cannot prevent U.S. policy makers from sitting down with their Chinese counterparts and saying, "We are shutting down our coal plants. When will you close yours?" That's essentially what happened between Barack Obama and Xi Jinping in November 2014.

But there's really only one mechanism for that to happen, and that's a price on carbon. By the end of 2014 that prospect seemed a bit closer, as the devastating economic effects of climate change became harder and harder to ignore; China inched toward a national carbon market; and even in the United States a de facto price on carbon, in the form of stiff federal penalties for greenhouse gas emissions, entered the realm of possibility.

In late 2014 the architecture firm Foster + Partners released its designs for its Battersea Roof Gardens, an undulating modernist apartment building that will, one day, house luxury flats topped by urban gardens with views of the old power station. The fortunate people who inhabit those flats will breathe air that was once full of mercury, sulfur, coal dust, and, of course, carbon. The air they breathe will be clean and carbon-free. That may be true, some day, for everyone, whether they live in the Ruhr Valley or southern Ohio or Shanxi Province. I won't live to see it. But my son may.

NOTES

PROLOGUE: CAPE FEAR

1. "Cape Fear Steam Plant," *SourceWatch*, http://www.sourcewatch.org/index.php/Cape_Fear_Steam_Plant#Plant_Data.
2. Investor-owned utilities are private, regulated companies that provide electricity to residential and business customers, often through power purchase agreements with local governments. They are distinguished from public utilities, which are usually owned by a government entity, such as a municipality.
3. Carl M. Cannon, "North Carolina's Changing Energy Mix," *RealClearPolitics*, October 10, 2014.
4. "Short Term Energy Outlook," U.S. Energy Information Administration, http://www.eia.gov/forecasts/steo/report/coal.cfm.
5. Ibid.
6. Rachel Cleetus et al, *Ripe for Retirement: The Case for Closing America's Costliest Coal Plants* (Washington, D.C.: Union of Concerned Scientists, November 2012).
7. Metin Celebi, Frank Graves, and Charles Russell, *Potential Coal Plant Retirements: 2012 Update* (Boston: The Brattle Group, October 2012).
8. "Clean Power Plan Proposed Rule," U.S. Environmental Protection Agency, http://www2.epa.gov/carbon-pollution-standards/clean-power-plan-proposed-rule.
9. "How Much Carbon Dioxide Is Produced When Different Fuels Are Burned?" U.S. Energy Information Administration, http://www.eia.gov/tools/faqs/faq.cfm?id=73&t=11.
10. "What Is U.S. Electricity Generation by Energy Source?" U.S. Energy Information Administration, http://www.eia.gov/tools/faqs/faq.cfm?id=427&t=3.
11. "Coal: Quick Facts," Center for Climate and Energy Solutions, http://www.c2es.org/energy/source/coal.

CHAPTER 1: THE TVA

1. Charlie Barnette, "In Search Of Col. James King's Iron Works," Bristol, Tenn-Va Collectible Bottles & History, March 1, 2010.
2. "Tennessee Eastman Kingsport Power Plant," *SourceWatch*, http://www.sourcewatch.org/index.php?title=Tennessee_Eastman_Kingsport_Power_Plant.
3. Hank Hayes, "Eastman Planning to Move from Coal to Gas," *Kingsport Times-News*, August 24, 2013.

4. Steven M. Neuse, "TVA at Age Fifty—Reflections and Retrospect," *Public Administration Review* 43, no. 6 (November–December 1983).

5. James Agee, "Where Did the Tennessee Valley Authority Come From?," *Fortune,* October 1933.

6. Ibid.

7. Leonard S. Hyman, *America's Electric Utilities: Past, Present and Future* (Reston, Virginia: Public Utility Reports, 1988), 74.

8. Merle Vincent, "Coal at the Cross-Roads," *Survey Graphic* 23, no. 4 (April 1934).

9. "Insull Drops Dead in a Paris Station," *The Montreal Gazette,* July 18, 1938.

10. Chris Martin, Mark Chediak, and Ken Wells, "Why the U.S. Power Grid's Days Are Numbered," *Bloomberg BusinessWeek,* August 22, 2013.

11. Peter Kind, "Disruptive Challenges: Financial Implications and Strategic Responses to a Changing Retail Electric Business," Edison Electric Institute, January 2013.

12. Martin, "Why the U.S. Power Grid's Days Are Numbered."

13. Jesse Berst, "Arizona Pushes Back Hard on Net Metering," *SmartGrid News* 14 (August 2013).

14. Martin, "Why the U.S. Power Grid's Days Are Numbered."

15. Shaila Dewan, "Hundreds of Coal Ash Dumps Lack Regulation," *New York Times,* January 7, 2009.

16. Shaila Dewan, "Water Supplies Tested After Tennessee Spill," *New York Times,* December 23, 2008.

17. "TVA Kingston Fossil Plant Coal Ash Spill," *SourceWatch,* August 24, 2012, http://www.sourcewatch.org/index.php?title=TVA_Kingston_Fossil_Plant_coal_ash_spill.

18. "North Carolina v. TVA," *SourceWatch,* August 25, 2011, http://www.sourcewatch.org/index.php/North_Carolina_v._TVA.

19. "Mitch McConnell Wants You to Know: Alison Grimes Believes in Climate Change," *Planet Experts,* September 20, 2013.

20. "McConnell Urges TVA to Keep Coal Power Plant Open," *PennEnergy,* October 28, 2013.

21. Erica Peterson, "Mitch McConnell Claims Influence in Coal Decision That Was Already Made," WFPL, November 28, 2013.

22. Pam Sohn, "TVA Trimming 1,000 Jobs and Delaying Some Capital Projects," *Chattanooga Times Free Press,* May 5, 2012.

23. "Electric Power Annual 2013," U.S. Energy Information Administration, http://www.eia.gov/electricity/annual/.

24. Older plants, which are more likely to be retired, tend to have smaller units in terms of capacity. Thus, while the average of all coal-fired units in the United States is about 262 megawatts, the average unit predicted to be shut down is closer to 150 megawatts. Daniel Bradley, "The EPA's Proposed Carbon Regulation: A Strategic Planning Perspective," Navigant, August 4, 2014.

25. Lincoln F. Pratson, Drew Haerer, and Dalia Patiño-Echeverri, "Fuel Prices, Emission Standards, and Generation Costs for Coal vs Natural Gas Power Plants," *Environmental Science & Technology* 47, no. 9 (2013): 4926–4933.

26. Barack Obama, "Remarks by the President on Climate Change," The White House, June 25, 2013.

27. Joe Uehlein, "Will Workers Be Left Behind in a Green Transition?," *The Nation,* May 18, 2009.

28. Ibid.

29. "Brayton Point Coal Plant: Operating at Our Expense," Coal Free Massachusetts, August 2013.

CHAPTER 2: KENTUCKY

1. William Martin, letter to Lyman Draper, June 1, 1842, http://www.oocities.org/gen josmartin/text8zz2.htm.

2. "Trends in U.S. Coal Mining 1923–2011," National Mining Association, June 2012, http://www.nma.org/pdf/c_trends_mining.pdf.

3. Ben A. Franklin, "Fatalism and Fatalities in the Appalachian Mines," *New York Times,* February 27, 1982.

4. "U.S. Coal Production by State & by Rank," National Mining Association, September 2014, http://www.nma.org/pdf/c_production_state_rank.pdf.

5. "Trends in U.S. Coal Mining 1923–2011," National Mining Association.

6. Erica Peterson, "Hollowed Mountains, Now Hollowed Towns: Coal in Eastern Kentucky," WFPL Louisville Public Radio, December 11, 2013.

7. Ibid.

8. "Gov. Beshear, Congressman Rogers Announce Citizen-Driven Summit to Envision Goals, Strategies," press release, Office of Rep. Hal Rogers, October 28, 2013, http://halrogers.house.gov/news/email/show.aspx?ID=PJI772SPIOQ4IQUUWZMF74O FQI.

9. Kris Maher and Tom McGinty, "Coal's Decline Hits Hardest in the Mines of Kentucky," *Wall Street Journal,* November 26, 2013.

10. "Stop Rx Drug Abuse Before It Starts," Office of the Attorney General, Kentucky, http://ag.ky.gov/rxabuse/Pages/default.aspx.

11. Ibid.

12. Attorney General Jack Conway, Testimony for the Committee on Energy and Commerce/Subcommittee on Commerce, Manufacturing, and Trade, March 1, 2012, http://ag.ky.gov/pdf_news/22712prescriptiondrugcongressionaltestimony.pdf.

13. "Barr Questions Treasury Secretary Jack Lew about Coal," Office of Congressman Andy Barr, December 13, 2013, http://barr.house.gov/media-center/videos/barr-ques tions-treasury-secretary-jack-lew-about-coal.

14. Ibid.

15. Matthew Daly, "EPA Denies Politics Delayed Pollution Rules," Associated Press, January 16, 2014.

16. Peterson, "Hollowed Mountains, Now Hollowed Towns."

17. Ibid.

18. Ronald D. Eller, *Uneven Ground: Appalachia Since 1945* (Lexington: University Press of Kentucky, 2008).

19. Ibid.

20. B. H. Luebke and John Fraser Hart, "Migration from a Southern Appalachian Community," *Land Economics* 34, no. 1 (1958).

21. Eller, *Uneven Ground.*

22. "Kentucky Darby, LLC Darby Mine No. 1 Explosion," *Mine Disasters,* U.S. Mine Rescue Association, accessed May 11, 2014, http://www.usmra.com/saxsewell/darby.htm.

23. Carl Shoupe, "Tired of Big Coal Telling Our Pols What to Do," *Lexington Herald-Leader,* May 7, 2014.

CHAPTER 3: WEST VIRGINIA

1. John F. Kennedy, "Remarks of Senator John F. Kennedy at Morgantown, West Virginia, April 18, 1960," John F. Kennedy Presidential Library & Museum.
2. Ibid.
3. Ibid.
4. Ibid.
5. Ibid.
6. Senator Jay Rockefeller, "Statement on Inhofe Resolution of Disapproval," Office of Senator Jay Rockefeller, June 20, 2012, http://www.rockefeller.senate.gov/public /index.cfm/floor-statements?ID=23709dde-73bc-4377-ac1a-0bc051d83449.
7. Ibid.
8. Ibid.
9. "Trends in U.S. Coal Mining 1923–2011," National Mining Association, June 2012, http://www.nma.org/pdf/c_trends_mining.pdf.
10. "The Coal Severance Tax," West Virginia Coal Association, http://www.wvcoal .com/taxes.html.
11. Philip Bump, "Bathtub Photo Lands Coal Activist in Child-porn Hot Water," *Grist,* June 4, 2012.
12. Ken Ward Jr., "Reactions to the EPA Announcement," Coal Tattoo, *West Virginia Gazette,* June 2, 2014.
13. "United States of America v. Donald Blankenship." U.S. District Court for the Southern District of West Virginia, Charleston Grand Jury 2014, November 12, 2014, http://www.wvgazette.com/assets/PDF/CH62291113.pdf.
14. Ibid.
15. "Sen. Jay Rockefeller Releases Statement on Indictment on Don Blankenship," WVVA, November 13, 2014.
16. Anya Litvak, "Murray CEO Predicts Doom for Coal, Says Own Company Will Survive," *Pittsburgh Post-Gazette,* September 23, 2014.
17. Steven Mufson, "After Obama Reelection, Murray Energy CEO Reads Prayer, Announces Layoffs," *Washington Post,* November 9, 2012.
18. Steve Horn, "Coal Baron and Major Ken Cuccinelli Campaign Donor Sues Blogger for Defamation, Invasion of Privacy," DeSmogBlog (blog), November 3, 2013, http://www.desmogblog.com/2013/11/03/coal-baron-major-ken-cuccinelli-cam paign-donor-sues-blogger-defamation-invasion-privacy.
19. James R. Carroll, "Coal Operator Lashes out over New Dust Rules," *The Courier-Journal,* April 26, 2014.
20. Rick Shrum, "Murray Energy Ends Medical Coverage for Salaried Consol Retirees," *Observer-Reporter,* April 16, 2014.
21. "Murray Energy Contributions," OpenSecrets.org, Center for Responsive Politics, http://www.opensecrets.org/orgs/summary.php?id=D000022123&cycle=A.
22. Ken Ward Jr., "Former Foreman Sues Murray Energy over Firing," *The Charleston Gazette,* September 9, 2014.
23. Neela Banerjee, "Ohio Miners Say They Were Forced to Attend Romney Rally," *Los Angeles Times,* August 29, 2012.
24. Renée Johnson, "Hemp as an Agricultural Commodity," Congressional Research Service, Library of Congress, February 14, 2014, http://www.votehemp.com/PDF /RL32725-20140214.pdf.
25. Katherine Andrews, "Hemp: An Energy Crop to Transform Kentucky and West Virginia," Patriot BioEnergy, 2014, http://www.patriotbioenergy.com/data/uploads

/whitepaper/whitepaper_an-energy-crop-to-transform-kentucky-and-west-virginia
.pdf.

26. Stuart Hammer, *NewsWatch,* WOAY Television Online, May 21, 2014.

27. Jenny Hopkinson, "Mitch McConnell High on Hemp Provision in Farm Bill," *Politico,* January 29, 2014.

28. Rob Parks, "Appalachian Heartbreak: Time to End Mountaintop Removal Coal Mining," Natural Resources Defense Council, November 9, 2009, https://www.nrdc
.org/land/appalachian/files/appalachian.pdf.

29. Ibid.

30. Ibid.

31. "Economic Impacts of Mountaintop Removal," *Appalachian Voices,* accessed June 22, 2014.

32. Christopher Barton, Michael French, Songlin Fei, Kathryn Ward, Robert Paris, and Patrick Angel, "Appalachian Surface Mines Provide a Unique Setting for the Return of the American Chestnut," *Solutions Journal* 1, no. 4 (July 2010).

33. Ibid.

34. Tom Hamburger, "A Coal-Fired Crusade Helped Bring Crucial Victory to Candidate Bush," *The Wall Street Journal,* Dow Jones & Company, June 13, 2001.

35. Ibid.

36. Tom Miller, "Absentees Dominate Land Ownership" *Huntington Herald-Advertiser,* December 1974.

37. Thomas Frank, *What's the Matter with Kansas?: How Conservatives Won the Heart of America* (New York: Metropolitan, 2004).

38. Nick Mullins, "The War on Coal," The Thoughtful Miner (blog), September 20, 2011, http://www.thethoughtfulcoalminer.com/2011/09/war-on-coal.html.

CHAPTER 4: WYOMING

1. Dudley Gardner and Verla Flores, *Forgotten Frontier: A History of Wyoming Coal Mining* (Boulder, CO.: Westview Press, 1989), 4.

2. Ibid.

3. Ibid.

4. Chris Probst, "Rock Springs, Wyoming," WyoHistory.org, http://www.wyohistory.org
/essays/rock-springs-wyoming.

5. T. A. Larson, *History of Wyoming* (Lincoln: University of Nebraska Press, 1978), 114.

6. Ibid.

7. ElDean V. Kohrs, "Social Consequences of Boom Growth in Wyoming," Rocky Mountain American Association for the Advancement of Science Meeting, Laramie, Wyoming, April 24–26, 1974, http://www.sublettewyo.com/archives/42/Social_Consequences_of_Boom_Growth_In_Wyoming_-_Kohrs[1].pdf.

8. Ibid.

9. "The Powder River Basin—A Root Contributor to Global Warming," *WildEarth Guardians,* http://www.wildearthguardians.org/site/PageServer?pagename=prioritie
s_climate_energy_coal_powder_river_global_warming.

10. Ibid.

11. William J. Nicolls, *The Story of American Coal* (New York: J. B. Lippincott, 1896), 33.

12. John McPhee, *Rising From the Plains* (New York: Farrar, Straus, Giroux, 1986), 7.

13. "Global Coal Demand Growth Slows Slightly, IEA Says in Latest 5-year Outlook," International Energy Agency, December 16, 2013, http://www.iea.org/newsroomand

events/pressreleases/2013/december/global-coal-demand-growth-slows-slightly-iea
-says-in-latest-5-year-outlook.html.

14. "International Energy Outlook 2014," U.S. Energy Information Administration, September 9, 2014, http://www.eia.gov/forecasts/ieo/pdf/0484(2014).pdf.

15. "Electricity Monthly Update," U.S. Energy Information Administration, October 24, 2014, http://www.eia.gov/electricity/monthly/update/.

16. "U.S. Per Capita Electricity Use By State In 2010," Energy Almanac, California Energy Commission, http://energyalmanac.ca.gov/electricity/us_per_capita_electricity -2010.html.

17. "China: Wood Mackenzie Says Thermal Coal Demand Will Reach Nearly 7btpa by 2030," Wood Mackenzie, June 4, 2013, http://www.woodmacresearch.com/cgi-bin /wmprod/portal/corp/corpPressDetail.jsp?oid=11324244.

18. "Gregory Boyce," *Forbes.*

19. "Statement of Peabody Energy on National Carbon Plan," Peabody Energy, June 25, 2013, http://www.peabodyenergy.com/Investor-News-Release-Details.aspx?nr=286.

20. Joseph Casey, "Patterns in U.S.-China Trade Since China's Accession to the World Trade Organization." U.S.-China Economic and Security Review Commission, November 2012, http://www.uscc.gov/sites/default/files/Research/US-China_TradePa tternsSinceChinasAccessiontotheWTO.pdf.

21. "Mayor McGinn: Coal Trains Would Significantly Increase Delays at Railroad Crossings," City of Seattle, accessed December 22, 2013, http://mayormcginn.se attle.gov/coal-trains-would-significantly-increase-delays-at-railroad-crossings/.

22. "Jobs and Economic Development," Alliance for Northwest Jobs and Exports, http:// createnwjobs.com/learn-more/jobs-and-economic-development/.

23. "Peabody Energy Chairman and CEO Greg Boyce Outlines 'Peabody Plan' to Eliminate Energy Poverty and Inequality," Peabody Energy, September 14, 2010, http:// www.prnewswire.com/news-releases/peabody-energy-chairman-and-ceo-greg-boyce -outlines-peabody-plan-to-eliminate-energy-poverty-and-inequality-102882079 .html.

24. "Petition for Reconsideration by Peabody Energy Company," U.S. Environmental Protection Agency, February 11, 2010, http://www.epa.gov/climatechange/Down loads/endangerment/Petition_for_Reconsideration_Peabody_Energy_Company .pdf.

25. Vic Svec, Peabody Energy, personal communication with the author.

26. Christian LeLong et al, "The window for thermal coal investment is closing," Goldman Sachs Commodities Research, July 24, 2013, http://d35brb9zkkbdsd .cloudfront.net/wp-content/uploads/2013/08/GS_Rocks__Ores_-_Thermal_Coal _July_2013.pdf.

27. Anthony Yuen et al, "The Unimaginable: Peak Coal in China," Citi Research, September 4, 2013, https://archive.org/stream/801597-citi-the-unimaginable-peak-coal -in-china/801597-citi-the-unimaginable-peak-coal-in-china_djvu.txt.

28. Jared Diamond, *Why Is Sex Fun: The Evolution of Human Sexuality* (New York: Basic Books, 1997).

29. "Bark Beetle Fact Sheet," Forest Service, U.S. Department of Agriculture, http:// www.fs.usda.gov/Internet/FSE_DOCUMENTS/stelprdb5337223.pdf; Christina Schmidt, "Forecast: North Platte Water Shortage Deepens," *Sheridan Press,* April 6, 2013; Harris Epstein, Johanna Wald, and John Smillie, "Undermined Promise: Reclamation and Enforcement of the Surface Mining Control and Reclamation Act 1977-2007," Natural Resources Defense Council and Western Organization of Resource Councils, 2007, http://www.worc.org/userfiles/file/SMCRA%20Report.pdf.

30. Neela Banerjee, "U.S. Loses Millions on Coal Leases, Inspector General Report Says," *Los Angeles Times,* June 12, 2013.
31. Jeremy Nichols, "Petition to Recertify the Powder River Basin as a Coal Production Region," *WildEarth Guardians,* November 23, 2009, http://www.wildearthguardians .org/support_docs/petition_powder_river_11-23-09.pdf.
32. Mark Squillace, "The Tragic Story of the Federal Coal Leasing Program," *Natural Resources & Environment* 27, vol. 3 (Winter 2013): 35.
33. Tom Sanzillo, *The Great Giveaway: An Analysis of the United States' Long-term Trend of Selling Federally-owned Coal for Less than Fair Market Value* (Washington: Institute for Energy Economics and Financial Analysis, 2012).
34. "Historical Coal Prices and Price Chart," *Investment Mine,* November 11, 2014.
35. Henning Gloystein, "Global Coal Prices Seen Dipping Further as China Seeks to Compete," Reuters, September 11, 2013.
36. "Permanent Wyoming Mineral Trust Fund," Wyoming Taxpayers Association, http://www.wyotax.org/PMTF.aspx.
37. "Wyoming not alone in abandoned mine land trust fund loss," Office of Representative Cynthia Lummis, July 12, 2012, http://lummis.house.gov/news/document single.aspx?DocumentID=303009.
38. Tarence Ray and Willie Davis, "Reclamation Money Could Fuel Recovery," *Daily Yonder,* August 11, 2013.
39. "High Plains Gasification-Advanced Technology Center Plans on Hold," Associated Press, July 29, 2011.
40. "Dry Fork Station," Basin Electric Power Cooperative, http://www.basinelectric .com/Electricity/Generation/Dry_Fork_Station/index.html.
41. "Dry Fork Power Plant," Powder River Basin Resource Council, http://www.pow derriverbasin.org/dry-fork-power-plant/.
42. Doug Hawes-Davis, director, *The Return,* High Plains Films, 2012.

CHAPTER 5: COLORADO

1. "Carpenter Ranch," The Nature Conservancy, http://www.nature.org/ourinitiatives /regions/northamerica/unitedstates/colorado/placesweprotect/carpenter-ranch.xml.
2. Erin Fenner, "Solar Array Likely Will Not Be Built in Craig," *Steamboat Pilot,* June 5, 2014.
3. Eli Stokols, "Environmentalists Flood EPA Hearing on Carbon Rules in Denver," FOX31 Denver, July 29, 2014.
4. Bob Beauprez, speech, Denver Coal Rally, July 29, 2014.
5. Lydia DePillis, "How Not to Shut down Coal Plants," *Washington Post,* July 24, 2013.
6. David O. Williams, "In Electricity War, It's Boulder vs. Xcel Energy," *Government Executive,* July 14, 2014.
7. Bob Greenlee, "People power," *Daily Camera,* April 6, 2008.
8. Ibid.
9. Erica Meltzer, "Boulder Leaders Divided on Xcel Agreement," *Daily Camera,* June 3, 2010.
10. "What's the Cost of Municipalization?," Xcel Energy, https://www.xcelenergy.com /staticfiles/xe/Corporate/Corporate PDFs/Final_11 08x10_Daily_Camera_oct2.pdf.
11. Meltzer, "Boulder Leaders Divided on Xcel Agreement."
12. Allen Best, "King Coal, Politics and the New Energy Economy," *ColoradoBiz,* September 1, 2010.

13. "Comanche Generating Station Unit 3," *SourceWatch,* http://www.sourcewatch.org /index.php?title=Comanche_Generating_Station_Unit_3.
14. Ibid.
15. Susan Osborne, "Guest Opinion: Xcel's Tactics to Hold onto City Are Unethical," *Daily Camera,* April 26, 2013.
16. Laura Snider, "Boulder Watchdog Denied Final Appeal to Participate in Xcel Cases at PUC," *Daily Camera,* October 20, 2011.
17. Leslie Glustrom, "Comanche 3: The Billion-Dollar Mistake," Clean Energy Action, March 31, 2010, http://cleanenergyaction.org/2010/03/31/comanche-3-the-billion -dollar-mistake/.
18. Mark Jaffe, "Analysis: Boulder Can Replace Xcel," *The Denver Post,* February 21, 2013.
19. Erica Meltzer, "Boulder Seeks Federal Ruling on Whether It Would Owe Xcel 'Stranded Costs,'" *Daily Camera,* May 17, 2013.
20. Upton Sinclair, *The Jungle* (New York: Barnes & Noble Classics Series, 2005).

CHAPTER 6: SHANGHAI

1. Chris Giles, "China Poised to Pass US as World's Leading Economic Power This Year," *Financial Times,* April 30, 2014.
2. Ibid.; "Coal Statistics," World Coal Association, http://www.worldcoal.org/resour ces/coal-statistics/.
3. "China Analysis," U.S. Energy Information Administration, http://www.eia.gov /countries/cab.cfm?fips=ch.
4. "Air Pollution in China," Facts and Details, http://factsanddetails.com/china/cat10 /sub66/item392.html.
5. Ailun Yang et al, "The True Cost of Coal—An Investigation Into Coal Ash in China," Greenpeace, August 2010.
6. "Developing Countries Subsidize Fossil Fuels, Artificially Lowering Prices," Institute for Energy Research, http://instituteforenergyresearch.org/analysis/developing -countries-subsidize-fossil-fuel-consumption-creating-artificially-lower-prices/.
7. Jos G. J. Olivier et al, "Global Trends in CO2 Emissions," European Commission Joint Research Centre, 2013.
8. Edward Wong, "In Step to Lower Carbon Emissions, China Will Place a Limit on Coal Use in 2020," *New York Times,* November 20, 2014.
9. Qu Jianwu, "Current situation of coal transportation and sales and the influence of a coal trading platform to the market and its development strategy" (Speech, Coaltrans Shanghai Conference, Shanghai, April 10, 2014).
10. Li Shuo and Lauri Myllyvirta, "The End of China's Coal Boom," Greenpeace, April 2014.
11. Denis Twitchett and John Fairbank, eds., *The Cambridge History of China,* vol. 1 (Cambridge: Cambridge University Press, 1986).
12. Barbara Freese, *Coal: A Human History* (Cambridge, MA.: Perseus Publishing, 2003).
13. Jonathan D. Spence, *The Search for Modern China* (New York: W. W. Norton, 1990).
14. Allison Rottman, *Resistance, Urban Style: The New Fourth Army and Shanghai, 1937–1945* (Berkeley: University of California Press, 2007).
15. Elizabeth Perry, *Anyuan: Mining China's Revolutionary Tradition* (Berkeley: University of California Press, 2012).
16. "Country Analysis Brief Overview: China," U.S. Energy Information Administration, http://www.eia.gov/countries/country-data.cfm?fips=CH#coal.

17. Nicholas Kristof, "China Plans Big Layoffs of Coal Mine Workers," *New York Times*, December 28, 1992.

18. Jonathan Fenby, *The Penguin History of Modern China* (London: Penguin, 2008).

19. Lu Yongxiang, ed., *Science Progress in China* (Atlanta, GA: Elsevier, 2006).

20. Gail Tverberg, "World Energy Consumption in Charts," *Our Finite World*, March 12, 2012.

21. Zhenya Liu, *Electric Power & Energy in China* (Hoboken, NJ: John Wiley & Sons, 2013).

22. LBNL China Energy Group, "Key China Energy Statistics 2012," Lawrence Berkeley National Laboratory, 2012, http://china.lbl.gov/sites/all/files/key-china-energy-statistics-2012-june-2012.pdf.

23. "Country Analysis Brief Overview: China," U.S. Energy Information Administration.

24. Ibid.

25. "China and Coal," *SourceWatch*, http://www.sourcewatch.org/index.php/China_and_coal.

26. Xu Nan and Zhang Chu, "What the World Is Getting Wrong about China and Climate Change," *China Dialogue*.

27. "China's Shenhua Asking Buyers for Help to Reduce Coal Stocks: Sources," *Latest News—Coal*, Platts, June 23, 2014.

28. "Revenue of China Shenhua Energy from 2010 to 2013," Statista, http://www.statista.com/statistics/227250/revenue-of-china-shenhua-energy/.

29. "Shenhua to Spend $602 Million on New Coal Loading Berths," *American Journal of Transportation*, May 29, 2012.

30. Vicky Validakis, "Shenhua Perseveres with NSW Coal Mine," *Mining Australia*, May 13, 2013; "Russia's Rostec, China's Shenhua to Invest $10 Bln in Coal Exploration," *Sputnik News*, September 7, 2014.

31. "Shenhua Energy's Annual Profit Drops 9 Pct on Weaker Coal Prices," Reuters, March 28, 2014.

32. Sarah McFarlane, "Coal Price Outlook Bearish until Mines Cut Supply," Reuters, November 4, 2014.

33. "China's Coal Consumption Growth Slows," English.news.cn, Xinhua, January 15, 2014.

34. "China's Coal Miners in Crisis," *South China Morning Post*, Reuters, April 2, 2014.

35. Bob Davis and William Kazer, "China's Economic Growth Slows to 7.7%," *Wall Street Journal*, January 20, 2014.

36. Yue Wang, "Caixin: Haixin Steel Debt Bigger Than Reported," *Forbes*, March 21, 2014.

37. Jeff Spross, "China Wants to Close 1,725 Coal Mines by the End of This Year," *ThinkProgress*, April 4, 2014.

38. Shuo and Myllyvirta, "The End of China's Coal Boom."

39. Christian Lelong, Jeffery Currie, Samantha Dart, and Philipp Koenig, "The window for thermal coal investment is closing," Goldman Sachs, July 24, 2013.

40. Anthony Yuen et al, "The Unimaginable: Peak Coal in China," Citi Research–Commodities, Citibank, September 4, 2013, https://archive.org/stream/801597-citi-the-unimaginable-peak-coal-in-china#page/n0/mode/2up.

41. "The End of China's Coal Boom," Greenpeace.

42. David Stanway, "China Approves Massive New Coal Capacity Despite Pollution Fears," Reuters.

43. Adam Sieminski, *International Energy Outlook 2013*, Center for Strategic and International Studies, July 25, 2013.

44. "China: Wood Mackenzie Says Thermal Coal Demand Will Reach Nearly 7btpa by 2030," Wood Mackenzie, June 4, 2013.
45. Ibid.
46. Yang Yufeng, "The Role of Coal in the Global Energy Mix" (Speech, Coaltrans Shanghai Conference, Shanghai, April 10, 2014).
47. Ibid.
48. Fan Baoying, "Latest mining technologies being introduced to improve efficiency of China's coal mines and reduce impact on the environment" (Speech, Coaltrans Shanghai Conference, Shanghai, April 10, 2014).
49. Will Freeman, "The Accuracy of China's 'Mass Incidents,'" *Financial Times*, March 2, 2010.
50. Zhang Youxi, "The Impact of Consolidation on China's Coal Industry" (Speech, Coaltrans Shanghai Conference, Shanghai, April 10, 2014).
51. Ibid.
52. William J. Kelly, "China's Plan to Clean Up Air in Cities Will Doom the Climate, Scientists Say," *Inside Climate News*, February 13, 2014.
53. Li Haofeng, "The Energy Outlook for China—What are the Key Factors Shaping Policy?" (Speech, Coaltrans Shanghai Conference, Shanghai, April 10, 2014).
54. Jamie Freed, "Future of Thermal Coal Cools," *Financial Review*, November 5, 2012.
55. Michael Taylor, "Indonesia Coal Companies in Panic, Closures Seen—Industry Group," Reuters, June 18, 2014.
56. Ibid.

CHAPTER 7: SHANXI PROVINCE

1. Didi K. Tatlow, "A Chinese Coal Baron Tumbles into Debt," Sinosphere (blog), *New York Times*, December 18, 2013.
2. "Coal Barons Return to Farmland," *Eastday*, April 10, 2014.
3. Ibid.
4. "Massive Plan to Monitor Shanxi Sinkholes," *China Daily*, March 11, 2014.
5. "The Pendulum Swings against the Pit," *Economist*, October 17, 2009.
6. Yanqin Ouyang, "Shanxi's No. 2 Graft Fighter Is Subject of Inquiry," *Caixin*, July 24, 2014.
7. Ibid.
8. Zhang Lu, "Jin Daoming Case Opens Corruption Wormhole," *Caijing*, June 3, 2014.
9. "Ta Shan Circulated Economic Park," Datong Coal Mine Group, http://english .dtcoalmine.com/101784/18955.html.
10. Li Yang, "Modern Life Arriving Slowly for Tongmei's Generations of Coal Miners," *China Daily*, May 1, 2014.
11. Ibid.
12. Armand Hammer and Neil Lyndon, *Hammer* (New York: G. P. Putnam's Sons, 1987).
13. W. Carl Kester and Richard Melnick, *The An Tai Bao Coal Mining Project* (Cambridge, MA: Harvard Business School, 1991).
14. Steve Weinberg, *Armand Hammer: The Untold Story* (London: Ebury Press, 1989).
15. David Barboza, "The Not So Good Earth," *New York Times*, June 23, 2006.
16. Jonathan D. Spence, *The Search for Modern China* (New York: Norton, 1990).
17. "Why Liquid Coal Is Not a Viable Option to Move America Beyond Oil," Natural Resources Defense Council, December 2011, http://www.nrdc.org/energy/files /liquidcoalnotviable_fs.pdf.

18. Jim Yardley, *Brave Dragons: A Chinese Basketball Team, an American Coach, and Two Cultures Clashing* (New York: Vintage, 2013).
19. Keith Bradsher, "Natural Gas Production Falls Short in China," *New York Times,* August 21, 2014.
20. "Shale Gas Revolution Stalls in China as Projected Output Halved," *Want China Times,* August 22, 2014.

CHAPTER 8: HANGZHOU

1. Frederick W. Mote, *Imperial China: 900–1800* (Cambridge: Harvard University Press, 2003).
2. Andy Brandl, "Leading The Way . . . into More Pollution?," *Photon Mix: Photography by Andy Brandl,* October 1, 2013.
3. "Zhejiang and Coal," *SourceWatch,* July 18, 2013, http://www.sourcewatch.org/in dex.php/Zhejiang_and_coal.
4. "It's Raining Coal in China's Hangzhou," *Epoch Times,* March 12, 2013.
5. Lucy Hornby, "China Waste Incinerator Protest Turns Violent," *Financial Times,* May 11, 2014.
6. Ibid.
7. Michael Nelles and Thomas Dorn, "Fuelling the Asian Dragon: WtE Challenges in China," *Waste Management World,* October 2012.
8. Hornby, "China Waste Incinerator Protest Turns Violent."
9. "Foreigners Wanted," *Economist,* August 28, 2014.
10. Li Yun and Din Ning, "Police Beat Back Coal Plant Protesters in Hainan Province," *The Epoch Times,* October 22, 2012.
11. "More Protests in Hainan as Government Revives Power Plant Plan," *Want China Times,* October 21, 2012.
12. "Hainan Travel Tourism Development Potential," World Travel & Tourism Council, July 1, 2012, http://www.wttc.org/focus/research-for-action/special-and-periodic-re ports/hainan-travel-tourism-development-potential/.
13. "Liu Futang's Verdict Causes Worry amongst China's Green Activists," *China Dialogue,* October 21, 2012.
14. Keith Richburg, "An Island's Dizzying, Troubling Growth," *Washington Post,* December 27, 2010.
15. "Ministry of Truth: Shenzhen Power Plant," *China Digital Times,* May 24, 2013.
16. Justin Guay, "Air Pollution Concerns Halt Enormous Coal Plant in China," *Huffington Post,* August 13, 2013.
17. "Improving Trustworthiness," *China Daily,* April 2, 2013.
18. Lauri Myllyvirta, "How Air Pollution Concerns Stopped a China Coal Power Project," *Greenpeace International,* August 14, 2013.
19. "In Asia Interviews TIMEs Hero Chinese Environmentalist Zhao Zhong," *In Asia,* Asia Foundation, October 24, 2012.
20. Austin Ramzy, "Heroes of the Environment 2009," *Time,* September 22, 2009.
21. Jake Schmidt, "China Leading the Clean Energy Race—Check out the Facts," Natural Resources Defense Council, March 25, 2010, http://switchboard.nrdc.org/blogs /jschmidt/china_leading_the_clean_energy_race.html.
22. He Jiankun and Nicholas Stern, "China's Climate Commitment," *Project Syndicate,* July 23, 2014.
23. "China Aims High for Carbon Market by 2020," *Sydney Morning Herald,* September 12, 2014.

24. It should be noted that many scholars of Mandarin dismiss this interpretation. See Victor Mair, "'Crisis' Does NOT Equal 'Danger' Plus 'Opportunity,'" Pinyin.info, September 1, 2009.

CHAPTER 9: OHIO

1. Dan Gearino, "Workers Sue AEP over Health Risks at Power Plant's Landfill," *Columbus Dispatch,* September 6, 2014.
2. "Cheshire, Ohio No More," *Cincinatti Enquirer,* May 7, 2002.
3. Eve Morgenstern, email to the author, October 5, 2014.
4. "Coal's Ticking Timebomb: Could Disaster Strike a Coal Ash Dump near You?," The Institute for Southern Studies, January 2009, http://www.southernstudies .org/2009/01/coals-ticking-timebomb-could-disaster-strike-a-coal-ash-dump-near -you.html.
5. Bob Cramer, "American Electric Reports Healthy Second-Quarter Results," *Bidness,* July 25, 2014.
6. Kitty McConnell, "Q&A with Nick Akins, American Electric Power," *Columbus CEO,* November 20, 2013.
7. "Samuel C. Randazzo," McNees Wallace & Nurick LLC, http://www.mwn.com /professionals/xprProfessionalDetailsMNW.aspx?xpST=ProfessionalDetail&profess ional=53&service=20.
8. Samantha Williams, "Ohio's Clean Energy Law Is Key to Unlocking Carbon Reductions," Natural Resources Defense Council, June 16, 2014, switchboard.nrdc.org /blogs/swilliams/ohios_clean_energy_law_is_key.html.
9. Deirdre Shesgreen and Maureen Groppe, "New EPA Power Plant Rules Affect Ohio More than Most," *Cincinnati Enquirer,* June 3, 2014.
10. Jay Miller, "As Electric Deregulation Takes Effect, FirstEnergy, AEP Drop Gloves," *Crain's Cleveland Business,* May 14, 2012.
11. "Federal Coal Subsidies," *SourceWatch,* CoalSwarm/Center for Media & Democracy, http://www.sourcewatch.org/index.php?title=Federal_coal_subsidies.
12. "Electric Power Monthly," U.S. Energy Information Administration, http://www .eia.gov/electricity/monthly/.
13. Trish Demeter, "Ohio House Public Utilities Committee Opponent Testimony— Am. Sub. Senate Bill 310," Ohio Environmental Council, May 27, 2014, http://www .theoec.org/publications/ohio-house-public-utilities-committee-opponent-testimo ny-%E2%80%93-am-sub-senate-bill-310-may.
14. Ibid.
15. Jeremy Pelzer, "Ohio Legislature Approves Two-year Freeze on Renewable Energy, Energy Efficiency Standards," *Plain Dealer,* May 28, 2014.
16. Disclosure: Steyer was a college classmate of mine, and he and I have been in communication during the writing of this book. Also, the company I work for, Navigant Research, has done work for Advanced Energy Economy.
17. Matt Sanctis, "Everpower Seeks Extension for Wind Farm," *Springfield News-Sun,* July 16, 2014.
18. Dan Burns and Lucia Mutikani, "U.S. Manufacturing Activity Holds near 4-1/2-year High," Reuters, September 23, 2014.
19. "Renewable Energy in Manufacturing," *IRENA,* June 2014.
20. Fred Discenzo, "Energy Savings in Manufacturing" (Speech, Northern Ohio Energy Management Conference, September 23, 2014), https://www.mecseminars.com

/sites/default/files/2014/september/9th-annual-northern-ohio-energy-management
-conference/2014-northern-ohio-energy-agenda.pdf.

21. "Delivered Energy Consumption by Sector," U.S. Energy Information Administration, December 16, 2013, http://www.eia.gov/forecasts/aeo/er/early_consumption
.cfm.

22. James Manyika, "Manufacturing the Future: The Next Era of Global Growth and Innovation," McKinsey & Company, November 2012, http://www.mckinsey.com
/insights/manufacturing/the_future_of_manufacturing.

23. "Manufacturing Sector Energy Use and Energy Intensity down since 2002," U.S. Energy Information Administration, March 19, 2013, http://www.eia.gov/pressroom
/releases/press383.cfm.

24. Jonathan Lesser, "Ohio's Energy Renaissance" (Speech, Northern Ohio Energy Management Conference, September 23, 2014), https://www.mecseminars.com
/sites/default/files/2014/september/9th-annual-northern-ohio-energy-management
-conference/2014-northern-ohio-energy-agenda.pdf.

25. Ibid.

26. "Public comment in support of FirstEnergy's ESP plan filed by R. Murray, Murray Energy Corporation," The Public Utilities Commission of Ohio, http://dis.puc
.state.oh.us/DocumentRecord.aspx?DocID=19906231-6c7f-423b-a27d-a121f3778
7c6.

27. Ibid.

28. Elizabeth Shogren, "5 Key Takeaways From the Latest Climate Change Report," *National Geographic,* November 2, 2014.

EPILOGUE: THE RUHR

1. Tim McDonnell, "This Town Was Almost Swallowed by a Coal Mine," *Mother Jones,* April 24, 2014.

2. Ibid.

3. Tino Andresen, "Coal Returns to German Utilities Replacing Lost Nuclear," Bloomberg.com, April 15, 2014.

4. Matthew Carr, "Rising German Coal Use Imperils European Emissions Deal," Bloomberg.com, June 20, 2014.

5. "Germany's Coal Binge," *Wall Street Journal,* September 24, 2014.

6. Ailun Yang and Yiyun Cui, "Global Coal Risk Assessment," World Resources Institute, November 2012, http://www.wri.org/sites/default/files/pdf/global_coal_risk
_assessment.pdf.

7. "The Unwelcome Renaissance," *Economist,* January 3, 2013.

8. "A Risky Proposition: The Financial Hazards of New Investments in Coal Plants," Union of Concerned Scientists, March 2011, http://www.ucsusa.org/clean_energy
/smart-energy-solutions/decrease-coal/financial-hazards-of-coal-plant-investments
.html#.VHYL8YvF_pU.

9. "NYC and NYS pension funds should divest coal stocks: A shrinking industry, weak upside, and wrong on climate change," Institute for Energy Economics and Financial Analysis, May 8, 2014, http://www.ieefa.org/report-nyc-and-nys-pension
-funds-should-divest-coal-stocks-a-shrinking-industry-weak-upside-and-wrong-on
-climate-change/.

10. James Legge, "Another Attempt at Redevelopment Kicks off at Battersea Power Station," *The Independent,* July 5, 2013.

11. "U.S. Coal Production, 2008–2014," U.S. Energy Information Administration, http://www.eia.gov/coal/production/quarterly/pdf/t1p01p1.pdf.

12. "May 2013 National Industry-Specific Occupational Employment and Wage Estimates—NAICS 212100—Coal Mining," U.S. Bureau of Labor Statistics, http://www.bls.gov/oes/CURRENT/naics4_212100.htm.

13. Karen Lee Ziner, "Sen. Sheldon Whitehouse Says There Are More U.S. Jobs in Solar Industry than Coal Mining," *Politifact*, July 6, 2014.

INDEX